THE LOG BOOK

*Getting the best from
your woodburning stove*

Will Rolls

Permanent Publications

Published by
Permanent Publications
Hyden House Ltd
The Sustainability Centre
East Meon
Hampshire GU32 1HR
England
Tel: 01730 823 311
Fax: 01730 823 322
Overseas: (international code +44 - 1730)
Email: enquiries@permaculture.co.uk
Web: www.permaculture.co.uk

Distributed in the USA by:
Chelsea Green Publishing Company, PO Box 428, White River Junction, VT 05001
www.chelseagreen.com

Designed by Two Plus George, www.TwoPlusGeorge.co.uk

Cover photographs:
front: Darren Pyke Photography
back: Wiktor Hasel/shutterstock.com

Printed in the UK by Bell & Bain Ltd, Thornliebank,
Glasgow

All paper from FSC certified mixed sources

The Forest Stewardship Council (FSC) is a non-profit international organisation
established to promote the responsible management of the world's forests.
Products carrying the FSC label are independently certified to assure consumers
that they come from forests that are managed to meet the social, economic and
ecological needs of present and future generations.

British Library Cataloguing-in-Publication Data
A catalogue record for this book is available from the British Library

ISBN 978 1 85623 157 2

Contents

Acknowledgements

So here we are, a second edition. I'm not sure I ever really thought it would happen. As always, I have a few people to thank:

Biggest thanks to my wife Lizzie, for the tolerance, encouragement and support.

To everyone who helped with the first edition and sent me comments, corrections and feedback for the second edition.

To the team at Permanent Publications for taking the book on, and being patient with a number of hardware failures at my end...

I. Introduction

WHAT IS A LOG STOVE?

Stoves are self-contained single-room heaters that are available to run on a wide variety of fuels. While log stoves that provide hot water for radiators and domestic use are available, the most common type of log stove is made from simple technology, requiring no electricity supply and heats only the room in which it is installed.

WHY USE ONE?

So, why use a log-burning stove at all? I'm often met with polite incomprehension when I suggest that wood may be a suitable alternative heating source. Many people I speak to see wood as only good for putting in open fires, largely for show rather than heat, or cutting down trees as actively bad for the environment. Neither of these need be true.

Security of supply

We live in an age of uncertainty. Environmental damage and the increasing scarcity of a wide range of resources are never far from our TV screens or newspapers. The fact is though, that wood, in most of the UK, is an under-used resource, which can offer significant benefits over traditional fossil fuels, in terms of the economics, the environment and security

1

of supply. Wood-fired stoves usually operate without any use of electricity and if you live in a remote area, the ability to heat independently from an outside source can quite literally be a lifesaver. Even city dwellers can appreciate knowing that if the boiler breaks down on Christmas day, they can still keep at least one room of the house warm.

The environment

I often encounter very well-meaning people who want to tell me just how bad for the environment cutting down trees, and then burning them, really is. Please don't misunderstand me, unsustainable forestry (deforestation) is worryingly common over much of the world. It has led to habitat decline and a loss of valuable species, as well as accelerating the problems presented by climate change. This is significantly different from what I'm trying to promote, for a number of reasons:

1. The forestry industry in the UK is one of the most tightly regulated in the world. If you want to fell trees totalling more than 5 cubic metres (m^3) in a calendar quarter, or sell more than $2m^3$, then you have a legal obligation to apply to the Forestry Commission for a felling licence. This will only be issued if what you are proposing is in keeping with good forestry practice (i.e. sustainable) as measured against the UK Forestry Standard. There are also regulations to protect endangered species and protected habitats. The UK Forestry Standard is based on internationally agreed sustainability criteria, which has led to FSC classifying UK woodland as 'low risk'. The Forestry Commission website,[†*] has lots of information on this if you are interested.

2. 'The wood that pays is the wood that stays'. Over recorded history, the main reason for the retention for woodlands has been the benefit they have offered to people in the local area in terms of useful products. This is primarily an economic reason. The most obvious way to guarantee the continuation of a resource is to make it economically viable to maintain it. In providing a local market for poor-quality timber, logs provide an important income for the rural economy and encourage active management of woodlands to retain their character.[‡]

3. The overwhelming majority of the landscape in the UK is a result of intensive interaction with humans over many hundreds of years. This has created the unique and varied character of the British landscape, and is particularly true of woodland. There are virtually no woodlands in the UK that have not been managed (i.e. felled) at some point. This has completely changed the character of the countryside and the nature of wildlife habitat, and has contributed to the valuable biodiversity that we see in British woodlands today.

Prior to active management, the natural landscape was composed of a complex patchwork of different habitats formed by interactions between natural events, such as floods, fires and storms; geology; topography; and climate. This pattern would naturally include many different woodland types, wetlands,

[†] www.forestry.gov.uk

[*] I have included many references throughout the book that I hope you will find useful. Unfortunately, many of them have very unhelpful internet addresses. To make things a little easier, links to all my references are available at: www.wrolls.co.uk/thelogbook2/reference

[‡] For more information see Rackham, O, *Trees and Woodland in the British Landscape*, Phoenix Press, London, 2001.

heaths and open spaces. Human manipulation of this landscape has changed and fragmented the prehistoric habitats and resulted in a complete change of the relationship between many species and their environment. Today, many woodlands are completely unmanaged plantations, this replicates a single (not very common) habitat type from the ancient pattern. Good forest management replicates the natural patchwork of habitats (albeit on a smaller scale) within a woodland. This has been shown to increase biodiversity within woodland habitats and develop the quality of our countryside for a wider range of species.*

Wood is a very carbon-lean fuel. Net carbon dioxide (CO_2) emissions are significantly lower when burning wood than burning any of the fossil fuels. This is because wood is composed of contemporary CO_2, which has been trapped (or sequestered) by photosynthesis, and the only net emission of CO_2 comes from processing and transport. In the meantime the woodland area has been replanted, allowing a continued CO_2 uptake. It is important to recognise that if we did not use wood in this way, a significant volume of poor-quality timber would remain un-harvested, or be left to rot in the woodland. While rotting, timber breaks down into CO_2 (and methane (CH_4) – which is a significantly worse greenhouse gas than CO_2). For more information on carbon balances of forestry and wood fuel use either have a look at the Biomass Energy Centre website[†] or see the Forest Research Note: *Forests, Carbon and Climate change: the UK contribution*.[‡]

Log stoves are not open fires

Unfortunately, the perception of log stoves is often that they are broadly equivalent to an open fire. Open fireplaces are hardly ever the most appropriate option for heating a home.

The majority of open fires in the UK have grates that were designed to burn coal, which makes it difficult to burn wood effectively. If you can get the wood to burn well, you will get a maximum of around 20 per cent of the calorific value of the wood radiated into your room, while the remaining 80 per cent will go up your chimney (often as unburned particulate matter). When the fire is not in use, the chimney continues to draw (suck air out of the room), and you will find that you have a strong convection current that allows all of the warm air from within the house to disappear into the sky – causing a net **loss** of heat.

On the other hand, there are a wide-range of different wood-fired heating boilers on the market which, unlike most stoves, can be connected directly to a central heating system with radiators. They can be designed to accept a range of different fuel types, from broken pallets and joinery off-cuts, to refined wood pellets, which allow for sophisticated control systems and automatic operation. While these systems may offer significant savings in running costs (particularly when compared to fuel oil) they still represent a big investment. The financial cost of installing biomass boilers is significantly higher than fossil-fuelled systems and they frequently need significantly more space for the boiler unit itself, as well as associated infrastructure, electricity supply and fuel storage.

---

* See the *Position Statement by Wildlife and Countryside Link on the Forestry Commission's Woodfuel Strategy for England,* available to download from www.wrolls.co.uk/wp-content/uploads/2012/07/Link_position_statement_Woodfuel_Strategy_03Jul09.pdf

† www.biomassenergycentre.org.uk

‡ Broadmeadow, M and Matthews, R, *Forests, Carbon and Climate change: the UK Contribution.* Edinburgh: Forest Research, Edinburgh, 2003, available to download from www.wrolls.co.uk/wp-content/uploads/2012/07/fcin048.pdf

Log stoves are a good middle way. They are a simpler, cheaper alternative to a full biomass heating system, with far fewer moving parts that can go wrong. They also have a far greater efficiency than traditional open fireplaces, and prevent the loss of heat up the chimney when not in use. When speaking to people who have installed wood stoves, I've found that they have effectively shortened their heating season for their main boiler by a couple of weeks at either end by using the stove as well. Depending on the level of insulation in the property, a well-run stove can also have a significant impact on the amount of fuel that is needed to run a central heating boiler throughout the winter.

WHAT THIS BOOK DOESN'T TELL YOU

Please note that this book **does not** tell you how to install a stove. I have provided a section on choosing a stove and finding an installer, but I assume that you will get it installed by a competent person in compliance with building regulations and good practice. If you've installed a stove yourself, or you have bought a property with a stove already installed, you should get it checked over by a qualified installer before you use it.

Incorrectly installed stoves, or stoves in rooms without sufficient ventilation, are unlikely to function efficiently and are prone to incomplete combustion. This is a significant hazard to health as it increases the proportion of solid particles, tars and other by-products released, many of which are toxic. Carbon monoxide (CO) is an odourless, colourless, poisonous gas that is released during combustion with poor ventilation. Around 50 people die in the UK annually due to poisoning by carbon monoxide* (from a variety of sources). **Please take this seriously**, every house should have smoke detectors as standard, but if you are running a woodburning appliance you must have an appropriate, properly installed, carbon monoxide detector in the same room as the appliance as a **bare minimum**. They are not expensive (I picked mine up for around £15 in a local hardware shop) and it's much easier to fit one than explain why you didn't to the coroner.

Less seriously, incomplete combustion is also going to shorten the life expectancy of your stove and flue liner, as the tar and soot deposits are acidic and will cause significant corrosion.

---

* According to the Department of Health, see www.dh.gov.uk/health/2011/11/co-poisoning

2. Combustion and the science of burning

BURNING

Setting fire to something is really very easy. Applying heat to fuel and air causes the fuel to react with the oxygen in the air and generate more heat. This leads to a chain reaction, which causes more more fuel to ignite and so on.

Remove one of these elements and the fire goes out. I first saw this triangle in a fire safety talk in primary school where the fire officer taught us why we put water on fires. In the case of most naturally occurring materials, water removes both heat and air from a fire, which either slows, or completely stops, the combustion reaction.

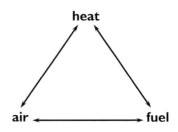

Figure 1. A Fire Triangle

So what do we do if we want the reaction to take place more efficiently, i.e. faster and more completely? First, let's look at a simple combustion reaction: oxygen and methane (natural gas).

$$2O_2 + CH_4 \rightarrow CO_2 + 2H_2O$$

oxygen + methane → carbon dioxide + water

It takes twice as much oxygen as methane to react completely, any less than this and there will be partially combusted gases floating about (usually carbon monoxide). Any more than this ratio, and the excess oxygen won't react; it has nothing to react with, instead it will only get in the way and have a slight cooling effect on the overall reaction.

This ratio of two oxygen molecules to one methane molecule is known to chemists as a stoichiometric ratio, this is just another way of describing an optimal mixture of fuel and air for complete combustion to occur. If fuel and oxygen are mixed in this ratio, after burning you should have none of either left – they will both have reacted completely.

This relationship assumes perfect mixing: both gases are freely available and mix completely. So what would happen if you tried to burn a liquid or solid fuel? The reaction can only take place where the fuel and oxygen come into contact, so a more accurate representation of our fire triangle would be something more like figure 2.

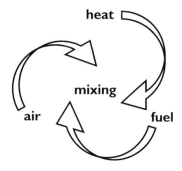

Figure 2. A more complex fire triangle

BURNING WOOD

Wood is a complicated fuel. Unlike methane or liquid fuels, the wood that goes in your stove is not refined or homogenised. It may be full of impurities and inconsistencies, and contains water; which is essentially inert. Wood properties vary between species, but also between trees of the same species growing in different locations, and even in different parts of the same tree. To make things even more complicated, different substances in wood melt and evaporate at different temperatures. There are ways of reducing and eliminating this complexity and running a wood stove very efficiently. One method is to refine wood into a pellet form, while the methods described here allow you to operate a log stove as efficiently as possible. In any case; around 80 per cent of the energy inherent in the wood comes from burning gaseous volatile compounds, which are driven off by the heat of combustion (secondary combustion), and only 20 per cent come from the solid 'char' which remains in the grate (primary combustion).

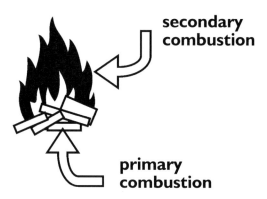

secondary combustion

primary combustion

Figure 3. Primary and secondary combustion

10

COMBUSTION VARIABLES

So, we've established the variables: fuel, air, heat and mixing. So what happens when we play with the different variables? How does the system react to these changes and what can we do about it?

Fuel

The value of fuel is measured in either joules (J) or kilowatt-hours (kWh – the standard unit provided on your utility bills). This calorific value is a measure of the amount of heat released per unit of fuel burnt completely – we want this value to be as high as possible.

There are a number of different variables that affect the calorific value. Understanding how variables such as moisture content and density affect calorific value, and controlling them effectively, is one of the best ways of increasing the efficiency of your stove.

■ *Moisture Content*

Wood is often described in terms of its moisture content. Wet wood doesn't burn well, and even if it does, energy will be used to evaporate the water within it. This has a very significant effect on the calorific value. In fact, even the term 'moisture' can be misleading; it implies that the wood is merely a bit damp, when in fact, if green wood wasn't so rigid, you'd be able to wring it out.* For every 1kg green

---

* There are two methods for calculating moisture content. I (and most fuel producers) use the wet basis measurement, it's certainly the easiest to grasp intuitively. If I hold a log in my hand that is at 50 per cent moisture content (wet basis) half of the weight is made up of water. If you want to know the difference between this and a dry basis sample look in Appendix 1.

11

log you add to a fire, you are also adding about a pint of water. Before your log begins to burn, all of that water has to evaporate off somewhere.

Table 1 shows the levels of seasoning you will commonly come across, with a description and indication of the moisture content to expect.

Table 1. Moisture contents at different levels of seasoning

	Description	Moisture content
Green	As it comes from the living tree	Usually around the 50% mark, though this is very variable
Part-seasoned	1 year of seasoning	Around 30–35%
Fully-seasoned	2 years of seasoning	Around 20–25%
Kiln-dry	Artificially dried to the sort of moisture content you'd expect to find in building timbers. Most wood sold as kindling should be in this category	Around 8–12%
Oven-dry	A theoretical limit. Wood that has been dried at 105°C for a number of hours. This is only really used when calculating moisture contents of wood samples	0%

The graph in figure 4 shows the average calorific value of wood at different moisture contents and shows what a big difference moisture content makes. Fully seasoned logs have approximately double the calorific value of the same weight of green wood.

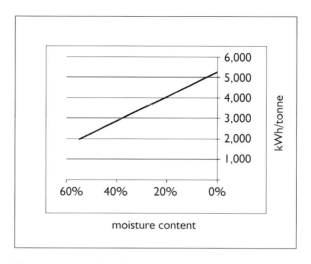

Figure 4. The effect of moisture content on calorific value

■ Density

Wood density has a negligible effect on the calorific value by **weight**. One tonne of poplar (a low density wood) is likely to have a very similar calorific value to a tonne of oak or beech (high density woods) at the same moisture content. However, if you're comparing species by **volume**, one tonne of oak will take up less space than a tonne of poplar, so you will need fewer logs to gain the same calorific value. Whether or not you think this is important largely depends on whether you are buying material by the tonne or by the m³ (or the 'load'). If you are using a particularly dense wood, you will probably find that you need to refill your stove less frequently, and you may find that you don't need as big a log pile.

13

■ Species

Perhaps surprisingly, the differences in chemical composition between species are relatively low.

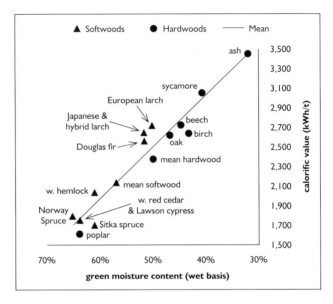

Figure 5. Calorific value of species by green weight

Most of the variation in calorific value is consistent with differing green moisture contents, not different species. You can tell the difference in calorific value between species by looking at the distance a species sits above or below the mean calorific value for a particular moisture content. Ultimately, the differences in the calorific value of seasoned wood are relatively small between species and are often outweighed by the effect of moisture content.

Another difference between species is that some have a tendency to spit when burnt (sweet chestnut, hawthorn and larch are some of the main culprits). This has, understand-ably, given them a bad name among people who like to burn on an open fire. In an enclosed stove though, spitting isn't really that significant, and I certainly wouldn't avoid species for that reason.

■ Hardwoods vs. softwoods

The distinction between hardwoods and softwoods is not really very helpful. This is one of a number of different clas-sifications of trees, and they all seek to put different species in categories based on specific characteristics.

Hardwoods

The term hardwood, has very little to do with the actual hardness of the wood (balsa is a hardwood). As a term it is usually used as equivalent to 'broadleaf', but what is actually being talked about is a group of trees that are from the *Angiospermae* (or flowering plants) group.

UK hardwoods tend to be denser than the softwoods grown in the UK. This is not always the case: poplar and willow, for instance, have a fairly low density.

Most UK hardwoods are deciduous (they lose their leaves in winter). A number of relatively common non-native broad-leaves, however are evergreen; *eucalyptus* for example.

Softwoods

Softwoods come from the plant family *Gymnospermae*, and are usually conifers (cone-bearing species). They tend to be less dense than UK hardwood species, but this is not always the case. Yew, for example, is a softwood that is significantly denser than many hardwoods.

Softwoods tend to have a higher proportion of resins and volatile oils. This can lead to a greater calorific value than hardwood species for the same weight of wood. Do be careful when burning softwoods wet, as the increased resin content may translate into a larger amount of tar on the inside of the flue. If you have wet softwoods, store them until they are dry, and then they are best added to fires that are already burning well and generating plenty of heat.

Softwoods are nearly all evergreen, there are however, a few notable exceptions such as larch.

The calorific value of timber is more or less constant across all species when measured by weight (at the same moisture content). If anything the softwoods do slightly better because of the increased resin content. The total variation in calorific values between all of the species (that I have data for*) is very low. The worst species has 95 per cent of the calorific value of the best species when measured by weight.

If we do the same calculation by volume (which takes into account the density of different species as in figure 6) the best species is oak, and the worst species is ash, which has 30

---

* Matthews, R and Mackie, E, *Forest Mensuration: A handbook for practitioners*, Forestry Commission, Edinburgh, 2006.

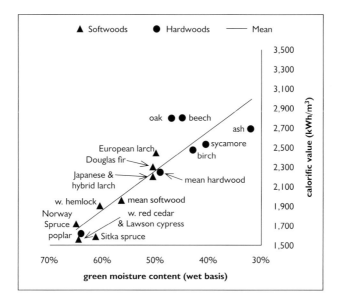

Figure 6. Calorific value of species by green volume

per cent less calorific value by volume. Both of them are hardwoods. I'm not suggesting that you shouldn't burn ash, it is often one of the best fire woods, but the reason for this is that it has an abnormally low green moisture content, not the fact that it's a hardwood.

■ Burning softwoods

There does seem to be a rumour floating around that burning softwoods is bad for stoves. **This is not the case** as long as the wood is properly dried. Burning any wet wood could damage your stove, because of an increase in condensation. If you think that you may have wetter wood (of any species) that you want to burn, make sure that you have got the fire good and hot before you put it on.

■ Poisonous plants

I have had a surprisingly large number of people asking me if it's safe to burn a whole range of species. Leyland cypress (*leylandii*) is a favourite, but I've had enquiries about laburnum, yew and other tree species. There are a lot of poisonous plant species out there, but most organic poisons are destroyed by exposure to the conditions present in efficient combustion (in fact, I can't think of any that aren't). So if the timber is dry enough to burn cleanly, and you have the air controls set correctly on your stove, for most species there is no problem.

To make sure, I have contacted the Poisons Information Service at the Centre for Economic Botany at Kew in London,* and the only specific species that they have any issues with is cherry laurel (*Prunus laurocerasus*). The main issue with this species is that it contains a number of compounds called *cyanogenic glycosides*. When any part of the plant structure (leave, branch or fruit stone) is broken, these react with the air to form cyanide gas. In outside conditions, this dissipates very quickly but in enclosed places, this can cause drowsiness, etc. There are anecdotal stories about people being made ill by the fumes when shredding the prunings in a closed garage or taking them to the local tip in a car with the windows closed, for example.

I have also come across anecdotal evidence that rhododendron can cause problems, though I believe that this has occurred when burning green foliage. The toxins present in rhododendron are harmful by ingestion, and it's not hard to imagine that burning this material green, on an open fire, would lead to a high concentration of the vaporised sap in the air. I do have to emphasise, though, that the risks associated

* www.kew.org/science/ecbot/poisonous-plants.html

with burning this material (after drying, and minus the foliage) in a properly installed and operated enclosed stove, do appear to be quite small.

On a personal note, I would be very happy if more people burned *leylandii*, it burns very nicely when it's properly seasoned...

■ Log size

Smaller sticks catch fire more easily than large logs. This is why we use smaller kindling to get the fire started. This effect is a result of a number of different causes:

Smaller diameter material is likely to be drier as the rate of drying is directly influenced by the distance moisture has to go before it is able to evaporate (i.e. the distance from the centre of a log to the surface).

A greater surface area to volume ratio on smaller sticks allows for more rapid mixing of the fuel and air.

Anecdotally, I know a number of people who make sure that every log has a split face, as this makes the logs easier to light. This may be because logs with split faces tend to dry more rapidly than round logs because the bark layer has been removed (so they are actually lighting the driest logs first), or it may be because the split faces have fibrous splintery material, which again increases the fuel surface area and catches more easily.

Smaller diameter material has a lower thermal mass so is able to reach ignition temperature more quickly. The thermal mass is also affected by the quantity of material that has to go through an endothermic process before it is able to ignite

– large logs will continue to absorb heat until they ignite, which takes longer.

The maximum size of log you should be using is around half of the volume of the firebox. Oversized material obstructs the heat and air from moving around inside the firebox and carries a lot of thermal mass with it. If the fire is not already very hot, you stand a good chance of the rate of burning slowing, if not stopping altogether.

There is no absolute minimum size of material, but using a large quantity of smaller twiggy material is going to leave you filling the stove pretty frequently. It is just not possible to get as much of this fuel into the stove at one time, and it will burn fast due to the effects I have just discussed. The exceptions to this are wood chip, wood pellets and some crumbly briquettes: these materials pack down tightly (or fall apart to sawdust, in the case of pellets and briquettes). This fine material restricts airflow to the fuel, which is likely to slow the combustion process. You can use this to manage the rate of combustion, but again, if overused this can put the fire out. Some stove manufacturers will provide a specially designed box that sits in the hearth specifically for burning wood pellets and wood chips.

■ Contamination

There are two basic forms of contamination: wood with 'extras', and things that look like wood (but aren't).

Extras

'Extras' can be (relatively) benign such as a bit of soil and grit, or water in wood that isn't quite dry enough. These things won't harm the stove (in moderation) but just don't help the burning process. The other category of extras, though, can be seriously bad news. Examples include wood with paint and preservative on it, chipboard, plywood, and (heaven forbid) tanalised timber. There are two issues associated with this sort of material: noxious flue gases and contaminated ash.

Halogenated organic compounds

Many man-made materials contain halogenated organic compounds, these usually include chlorine or similar chemicals, which are released during combustion and will lead to corrosive acidic elements in the flue gas (usually hydrochloric acid). These compounds are usually present in wood preservatives such as wood stain and varnish.

Heavy metals

Pigments in paints and preservatives often include heavy metal elements. These do not escape in the flue gas, but remain in particulate matter released and the bottom ash. These elements are also often the result of contamination on old industrial sites and can be drawn up by plants growing on site. Having these elements in the bottom ash greatly restricts your ability to make use of what is otherwise a useful material. The last thing you want to do with this sort of thing is put it on the compost. In particular, watch out for material that has been treated with the old tanalising process (green-stained, pressure-treated wood). The chemical used for this, until recently, contained chromium, copper and arsenic – not something that you want to expose your family to.

Driftwood

Wood that has been soaked in seawater is going to be full

of salt. Salt is going to increase the rate of corrosion inside your wood stove and increase the quantities of dioxins present in the smoke. It is best avoided altogether.

■ *Things that look like wood, but aren't*

There are a few products on the market that are being sold as briquettes to use in woodburning stoves that come from raw materials other than wood. While most of them are absolutely fine, a few aren't. Straw and *miscanthus* briquettes are pretty common, and while they probably won't extend the life of your stove (they both have relatively high chlorine and sulphur levels when compared to wood) they are unlikely to do any significant harm. However, do keep an eye out for briquettes that have been made with paper and cardboard. Printing inks often contain heavy metals which cause the problems I mentioned above, and the paper-making process uses significant amounts of sulphuric acid to remove the lignin from the wood (the substance that gives wood its rigidity). Most paper is more acidic than virgin wood when it reaches the end of its life.

Air

As far as woodburners are concerned, air comes in two different flavours: primary air, and secondary air. Primary air is fed in below the base of the fire and feeds the combustion of the solid char in the grate (primary combustion), secondary air is fed in above the fire bed and feeds combustion of volatile gases (secondary combustion).

Getting the total quantity of air going into the stove correct is important, but so is achieving the correct ratio between

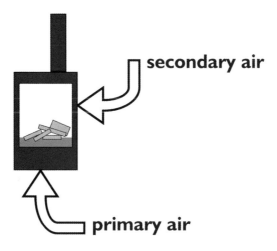

secondary air

primary air

Figure 7. Primary and secondary airflow

primary and secondary air. You should try and match the ratio of primary and secondary combustion with a ratio of between 1:4 and 1:5 primary to secondary.

■ *Controlling the air*

Controlling the airflow does change the rate of combustion within the stove, but it is not recommended as the primary method for regulating the stove. If you add too much air to a fire, you are liable to cause cooling rather than an increase in temperature (the draught sucks the heat up the chimney). Restricting airflow is equivalent to deliberately causing incomplete combustion. It does control the rate of the reaction, but is far from being the best way to do it. A much better method is to plan ahead and reduce the amount of unburned fuel in the system. Restricting airflow is also about the easiest way that I know of to get soot all over the glass panel in the front of your stove.

Heat

Heat is the objective, but it is also the means of sustaining the reaction. Ultimately, what you're looking for is as much heat as possible within the firebox, and this will lead to a warm room as a by-product. Estimates and quoted figures vary, but wood begins to burn at around 230°C and can reach temperatures of around 900°C.

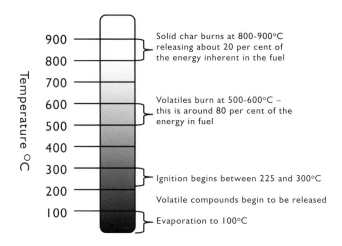

Figure 8. Temperatures achieved by burning wood

All of the processes involved in burning wood involve energy transfer. Up to ignition point, the process requires energy to be added (it is endothermic). Once ignition has been achieved, the process releases heat (it becomes exothermic); however, if the net energy balance throughout the fuel is still negative, the process may fail to become self-sustaining. Effectively, you need to release enough energy from the reaction to allow for the endothermic energy requirement in adjacent fuel, as well as achieving a sufficiently high temperature to allow ignition to take place. Wet fuel increases this energy requirement, as

24

the additional water in the wood has to evaporate before ignition can take place. Large logs can absorb significant amounts of heat in this way before they are ready to ignite.

Ideally what you want is an average temperature throughout the firebox in the middle of this scale. Too little heat, and the reaction becomes inefficient – you'll run the risk of fuel failing to ignite and incomplete combustion of the flue gas. Cool flue gases lead to condensation within the flue which will usually reduce its life expectancy. Gases should still be hotter than 120°C when they leave the flue.

In case you were wondering, it is technically possible to have too much heat in a combustion reaction (though in practice it hardly ever happens). At this point inefficiency is not an issue. The two main problems at this point are:

Clinkering

Silicates (sand to you and me) in the ash melt, from about 1,000°C, and then set onto the grate as clinker. This blocks up the air supply and leads to inefficient combustion in the future. It's important to note that a number of solid biomass fuels that come from non-woody sources (such as straw) have a lower ash melting temperature. This can lead to an increase in the amount of clinkering that you might encounter.

NO_x (Nitrogen Oxides)

It is possible to burn nitrogen out of the air, which is released into the flue gas. Compounds of nitrogen and oxygen (NO_x) are commonly produced as a result of burning coal, and other fossil fuels, and are one of the causes of acid rain. It is possible for burning wood to reach these temperatures, but is much more likely in a ultra-high efficiency pellet system rather than a conventional log stove.

Mixing

For efficient, complete combustion the fuel and air have to reach the right temperature and be in contact with each other. Effectively this means that you need to balance the amounts of the other variables to keep everything moving smoothly.

A failure to mix means that in localised parts of the stove the proportions of fuel, air and heat will be incorrect for efficient combustion. This may either be a result of the rate of mixing, or the extent of mixing.

Table 2. The effects of mixing on combustion

	Rate	Extent
Too great	Insufficient time is allowed for the transfer of heat between fuel and air, this slows the rate of combustion and reduces heat output. Too much mixing is usually associated with too much air entering the system. This tends to draw off heat up the flue and reduce the temperature. If the temperature drops too far, new fuel fails to ignite and the fire goes out.	If the fire is spread over too great an area, the effective temperature at any point within that fire risks being too low to achieve ignition of new fuel.
Too little	The areas where combustion is taking place are starved of **oxygen**. This reduces the rate of combustion and may lead to an increased production of CO and other products of incomplete combustion.	The areas where combustion is taking place are starved of **fuel**. This causes the temperature to drop. If the temperature drops too far, new fuel fails to ignite and the fire goes out.

3. Choosing and installing a stove

CHOOSING THE STOVE

Size

Specifying the size of a stove is (unfortunately) often more of an art than a science. There are a number of issues around the size of a system and it can be more complicated than you'd think.

There are a number of guides online for estimating how big a stove you'll need, but most of them seem to be little more than rules of thumb. There really is no substitute for discussing the project with an experienced professional, as the appropriate size of stove will vary depending on other heat sources, what the building is made of, how exposed the building is, what part of the country you live in, and the degree of insulation and double glazing etc.

Some of the things to bear in mind are:

■ *How big is the fireplace?*

Yes, I know, this is pretty obvious. However, the size of fireplace doesn't really give you much more than an upper limit for the size of stove. **Don't just automatically get the biggest stove that will fit in the space**. This is for two reasons: 1) because you could easily end up with a living room like a sauna, and 2) because using a stove continually turned down to its lowest settings is likely to cause you all sorts of problems with soot and tar production.

■ *How big is the door?*

Another obvious point. It's not going to have a huge impact on the running of the system, unless you have to cut every log you buy in half before you use it (which would certainly lead to reduced use in my house.) While some log suppliers will offer logs cut to a specific size, I would suggest that you'd have to have a very clear reason why you wanted to install a stove that needed a log shorter than about 10 inches (25 cm). This is a fairly average length for suppliers in the UK which will mean that you're able to shop around between suppliers to get a good price later.

■ *What's the rated output?*

It is rare to find a stove coming from a reputable manufacturer that doesn't have a quoted output rating. This is measured in kilowatts (kW) and gives an indication of the maximum output of heat from an appliance.* However, you should bear in mind that the rated output is a fairly flexible number, as it is basically equal to whatever the stove manufacturer can achieve with an optimum refuelling interval and good quality fuel (similar to a MPG quoted for cars). Again, don't automatically get the highest output you can afford (or that will fit), as you'll probably run into the problems listed above.

---

* A 1kW stove burning at full output for one hour, will generate 1 kilowatt-hour (kWh) of heat – it's useful to know this for calculating the amount of fuel you'll need (covered later on).

Features

There are a wide-range of different features in modern wood-burning stoves. These features are often designed to improve the efficiency of the appliance, but can also improve the appearance and ease of use.

■ Primary and Secondary Air

Any stove that you are using to burn wood should have both primary and secondary airflows. Occasionally these are integrated to a single draught control, which controls all the air reaching the combustion zone. It's easy to see why the stove designer would choose to go down this route – it restricts the amount an inexperienced user can play with the settings. If the draughts are permanently set at 1:5 primary to secondary, then it is impossible to fully close the secondary air; even when the primary air is completely cut off.

In my opinion, you really should ensure that any new stove you buy has independently controllable primary and secondary airflows, as this allows a much greater range of fine-tuning. Don't worry if it is impossible to close the secondary air vent fully, as there are no circumstances when you would need to do this during proper stove operation anyway.

■ Recirculation

Most modern stoves are designed to re-circulate air through the system. This allows for a more efficient heat transfer, and more efficient mixing of fuel and air during secondary combustion. Effectively this process involves passing the flue gas through the combustion zone to ensure that any particulate matter and other vapours have been completely burnt, reducing smoke, tar and fly-ash production.

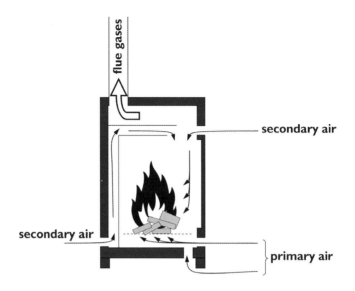

Figure 9. Primary and secondary air

▪ Pre-heating

Many stoves use a hollow door system. This uses a secondary airflow to insulate the stove door (which allows you to open the stove and add more fuel easily) while increasing the temperature of the air coming into the firebox. Heating the air going into a fire speeds up the combustion reaction by reducing the time needed to transfer heat from the on-going reaction to newer air and fuel.

▪ Air-wash

Air-wash systems are pretty common in glass-fronted stoves. This is a type of pre-heating mechanism that draws clean secondary air over the glass face of the door. In addition to the benefits associated with pre-heating, this keeps the glass in the door (relatively) clear of soot so that you can enjoy looking at the fire.

■ *Back Boilers*

There are some hybrid stoves on the market, which combine the room-heating function of a generic stove with the ability to provide domestic hot water and sometimes a small number of radiators. These can be tricky to operate well, as the water in the heating system can have the effect of cooling the firebox, providing a perfect surface for water in the flue gases to condense on to. There are ways of getting round this: if you're interested, I'd recommend getting hold of a copy of *Home Heating with Wood* by Chris Laughton* which covers these techniques in much greater detail.

■ *Direct air ventilation*

All stoves require a good source of air to burn efficiently. It is fairly common to draw this from the air in the room. However, to ensure that there is no build up of carbon monoxide or reduction of breathable oxygen in the room it is often necessary to add additional ventilation. Many older houses have enough unintentional holes and draughts that additional ventilation is not necessary, but if the stove is going into a newer building which is more effectively sealed, or if the stove is particularly large and will need a large air supply, additional ventilation may be required. This leads to an obvious drawback, as cold air is drawn into the room from outside to replace hot air going up the chimney (a very similar problem to the effect of having an open fire). Many newer

* Laughton, C, *Home Heating with Wood*, Centre for Alternative Technology Publications, Machynlleth, 2006.

stoves are designed to overcome this problem by drawing in air directly from outside into the firebox. This air does not circulate into the room and does not cause draughts. Direct air is only really feasible if you are installing the appliance on an exterior wall.

■ Multi-fuel stoves

There are a wide range of different stoves on the market that are sold as 'multi-fuel' meaning that they can accept a range of different solid fuel types. This is a compromise between the requirements of coal and wood (which means that it is optimised for neither).

To burn wood well on a multi-fuel stove, you should make sure that you are able to adjust the primary and secondary air. Since the insulating effect of ash is helpful for maintaining a wood fire, you may want to empty the ash less frequently or even blank off the grate completely with a metal plate.

If you want to burn wood and coal on the same fire, you will need to try and find a happy medium between the requirements for the two fuel types, though you do run the risk of getting the worst of both worlds.

The difference between coal and wood

At the fundamental level, growing vegetation and coal are composed of the same elements. In practise, however, there are significant differences:

Coal is fossilised vegetation, but this includes the whole plant (as opposed to just the solid, woody component) and a wide range of different contaminants including grit, soil, leaves, volcanic ash, small animals, etc. The net effect of this is an increase in the amount of sulphur and other unpleasant flue gases, and an increase in the solid matter left behind in the grate when combustion has stopped. This solid char has a higher proportion of heavy metal contaminants, and is certainly not something that you'd want to be putting on the garden. You will also find that there is significantly more of it than you would get from an equivalent amount of wood.

This difference in composition is also reflected in differences in combustion characteristics. Coal burns mainly as a solid lump, rather than as volatile compounds released during the reaction. This means that the proportions of primary and secondary air are very different to that needed by logs. As a rough rule of thumb, coal needs the opposite proportions of primary and secondary air (around 80 per cent primary and 20 per cent secondary), though many coal-fired systems are designed to do without a secondary feed altogether.

Table 3: Coal vs. wood

	Coal	Wood
Primary air	High requirement, burns best on an open grate allowing a draught unimpeded by bottom ash	Low requirement, burns best on a bed of ash which insulates the reaction
Secondary air	Low requirement, is sometimes eliminated altogether, burning residual pre-heated oxygen that has passed through the fire bed from the primary air feed	High requirement, most heat comes from volatile compounds released during combustion
Burning duration	Longer	Shorter
Flue gases	Greater proportion of sulphur and chlorine, burns hotter and is more likely to produce NO_x (corrosive components of acid rain) Reintroduces **fossil** carbon into the atmosphere	Low proportion of sulphur and chlorine, rarely burns hot enough to generate NO_x Reintroduces **contemporary** carbon into the atmosphere
Ash content	High silicate content leaves lots of solid ash which needs to go to landfill	Low silicate content, ash is non-hazardous and can be used as a soil additive (among other things)
Burning temperature	Higher	Lower

Flues and Chimneys

■ New flues

If you don't have an existing fireplace, then you will also need to install a flue. These should be made of stainless steel and insulated where necessary. Insulation is needed to maintain a high temperature in the flue gas when it exits the building, and also to make sure that the exterior of the flue remains cool enough that nothing gets damaged. It isn't always necessary to insulate the entire length of the flue, and uninsulated flues can provide a valuable additional source of radiant heat. If you're not sure about what sort of flue you should get and how it should be installed, first, check with an installer, and then if you want to make really sure, check the building regulations (part J). Do bear in mind that you might also need to get planning permission before putting up a new flue.

■ Using existing flues / chimneys

Where there is an existing chimney, you'll almost certainly want to get it lined. There are a number of reasons why this is a good idea:

Sealing the chimney

Many chimneys in older properties have never been lined and have exposed brickwork inside. Cracks and gaps in this sort of masonry will leak flue gasses and can cause significant air quality issues within the house. This kind of broken surface is also ideal for accumulating soot and tar deposits – increasing the likelihood of a chimney fire.

Constant diameter

Chimney diameter varies, and this has a significant impact on the draught. Larger diameter chimneys will tend to operate at lower pressure which will tend to cool the flue gasses more quickly and move them out of the building more slowly (leading to more condensation). Installing a liner will ensure that the flue is at the correct diameter for your stove though its entire length, allowing the stove to operate as designed.

Insulation

Insulation and thermal mass in buildings varies. Significant temperature losses in the flue will (again) lead to condensation and tar build up, increasing the risk of a chimney fire and corrosion to the register plate on your stove. A properly installed liner should be packed with insulation that keeps the flue gasses hot enough to exit the building before condensation occurs.

■ *Liners from gas appliances*

It's worth mentioning that you will certainly need to re-line a flue if you have had it lined for a gas appliance. The specification of flue for a woodburner is significantly more robust than the light-duty version used by a gas system that will corrode and rot away very quickly.

■ *Cowls*

Chimney cowls (or hoods) are not always essential for woodburning appliances, though they can be a useful tool for correcting some specific problems. If you are experiencing problems with rainwater leaking into your flue, birds nesting on the chimney, or serious down-draughts; you may find that these can be solved by the addition of a cowl. The units themselves are often relatively cheap (in the region of £50) but you may find that you spend quite a bit more than that to pay someone to fit it for you. It's also worth bearing in mind that adding the wrong cowl design, or adding a cowl unnecessarily, could actually create more problems than it solves by interfering with the draught in your flue.

Accessories

■ *Flue Temperature gauges*

These gauges are designed to fix to single skin stainless steel flues,* or on the side of a stove, and measure the temperature of the surface. Most have a magnet to hold them in place, and many are also designed to wire in place for a bit of added security. The idea is that they provide a handy guide for checking the temperature of your stove to ensure a good degree of combustion and to make sure that all flue gasses are well above 100°C before leaving the flue.

Temperature gauges will only ever give you an **indication** of temperature rather than a definitive reading. The temperature reading will vary depending on the temperature of the surrounding room, and I have even heard accounts of the reading being affected by the flue colour (dark colours radiating more effectively than bare steel). Most gauges will come with a gradation in degrees (usually in Fahrenheit and Centigrade) as well as a guide to show if the stove is too hot, too cold or just right. Do bear in mind though, that there doesn't seem to be a good consensus on where these brackets lie, and different gauges will be marked up differently. It is probably safest to accept that you won't know the exact temperature, and practise until you have a good idea of the range of readings best for your stove. You can expect a temperature gauge to cost in the region of £15-20 at the time of writing.

---

* Not insulated double skinned flues, for obvious reasons.

■ *Heat fans*

Heat fans sit on top of a stove and work based on the temperature difference between the base of the fan and a heat sink at the top. They are either based around an electric motor powered by a thermocouple, or a rather arcane bit of technology called a Stirling engine (an external combustion engine). The objective is to circulate hot air from immediately next to the stove around the room. This dramatically changes how the heat is transferred; from a system reliant primarily on radiated heat (with convection currents drawing warm air towards the ceiling), to a system that blows warm air into the room as well. It's not hard to see what a difference that this makes, air is a particularly good insulating agent, and moving this warm air around will prevent it insulating the stove, and reduce the tendency of warmer air to sit near the ceiling. Heat fans are particularly useful if you find that you are generating more heat than you need in the room with a stove, and other areas of the house are too cold.

A couple of words of caution, heat fans usually require a surface with a minimum temperature of around 60°C to operate. If your stove is inset tightly into a fireplace, or has a particularly well-insulated top surface, it's unlikely that you'll be able to get one of these to work. The other thing to bear in mind is that they can be quite delicate. I would certainly think twice before leaving one of these out in a room with either children or pets (particularly as they are usually quite expensive – £70-100).

- *Fireguards*

Please bear in mind that fireguards don't just apply to people with small children, a whole range of furnishings, papers, etc. can be a fire hazard if they fall onto a hot surface. The building regulations specify a minimum distance between a stove and any flammable material (a hearth, basically) and it's always good practice to make sure that it is clear before lighting the stove. The easiest way to ensure this in practice is to put in a fireguard. Don't be tempted to get an old fashioned 'spark guard', they are cheaper, but sparks probably won't be an issue for a stove and they are very unstable and easy to knock over. For preference, what you're after is something quite sturdy that can be properly fastened to a wall (some will also screw to the floor.) Prices vary, depending on how fancy you want it to look, but somewhere between £25 and £50 seems to be about average. Fireguards also age pretty well, so you might be able to pick up a cheaper one second hand.

FINDING AN INSTALLER

If you're looking to get a stove installed you want to find an installer who knows the business. You can install a stove yourself, but you really do need to check that you're complying with all the appropriate regulations. Apart from being good practice, incorrectly fitted stoves are a serious hazard, and can be a threat to the health of the occupants of the house. Personally, while I might consider putting in my own stove, I would be absolutely certain to get it checked by a professional before using it.

There are a few methods for tracking down an installer:

HETAS

HETAS is 'the official body recognised by Government to approve biomass and solid fuel domestic heating appliances, fuels and services including the registration of competent installers and servicing businesses'. Broadly speaking, they are the solid fuel equivalent of Gas-Safe (CORGI). While there is no legal obligation to use a HETAS trained and registered installer, they do have an extensive list of companies who have received the proper training (see www.hetas.co.uk).

The Stove Industry Alliance

The SIA is the trade organisation for stove installers in the UK. They have a list of their members on their website: www.stoveindustryalliance.com

Sellers/manufacturers

Many sales showrooms and stove manufacturers have good relationships with particular installers, or may install themselves. It's certainly worth checking – the Yellow Pages is the best place to find details.

Word of mouth

It's often better to find an installer by recommendation. If you have friends who have just had a stove installed, see if they got good service, and whether they are happy. You could also buy them a copy of this book!

REGULATIONS

Smoke control areas

As part of the Clean Air Act (1993), if you installing a stove within a smoke control area, you may only burn an approved smokeless fuel, unless you are using an exempt appliance. Smokeless fuels include Coalite and similar solid fuels, gas and anthracite. **No form of wood or other biomass is an approved fuel.** This means that you will need to install an exempt appliance if you want to burn logs within a smoke control area. There is a list of approved appliances on the DEFRA website at: http://smokecontrol.defra.gov.uk/appliances.php

If you have surplus fuel that you wish to sell, you should also be aware that if you knowingly sell an unapproved fuel for use in a smoke control area in an appliance that isn't exempt, you are committing an offence. However, there is no legal requirement for you to inspect a stove before you sell fuel to a customer and if a customer collects the fuel direct from you, then you have no realistic method of checking what equipment they are burning it in (though if you were selling fuel to a friend, you probably wouldn't be able to claim ignorance).

If you have questions about smoke control areas, the best people to speak to are the environmental health department at your local council.

Building Regulations

The Building Regs impose minimum standards for the design and construction of buildings. They cover health and safety for people in and around the buildings and energy efficiency. Part J covers combustion appliances and cover issues such as fire safety, provision of ventilation and requirements for flue design and dimensions.*

Planning Permission

Planning permission is not normally needed when installing a log stove in a house if the work is all internal. If the installation requires a flue outside, you should check with your local planning department, particularly if you live within a national park or conservation area.[†]

COSTS

Unfortunately, there are no hard and fast figures to give a good indication of cost, though according to *Which?*[‡] the going rate for an installed stove is around £1,500. This figure does look a little low to me, and I suspect that if you include the work required to fit a flue liner you may find that you spend a bit more than this.

A note on grants

While grants and subsidies for biomass boilers are available, they usually explicitly exclude stoves (usually on the grounds of efficiency, or because of the difficulties of measuring radiant heat produced). In terms of finding support, you may find that your most cost effective route would be to talk to an advisor about reducing your heat demand [‡‡] by installing insulation or other measures.

* www.planningportal.gov.uk/buildingregulations/approveddocuments
† More details are available here www.planningportal.gov.uk
‡ www.which.co.uk/energy/creating-an-energy-saving-home/guides/wood-heating-systems/how-to-buy-a-woodburning-stove
‡‡ The Energy Saving Trust is a good starting point: www.energysavingtrust.org.uk

4. Good stove operation

The best way to burn wood cleanly is fast and hot. This maximises the amount of heat produced and minimises any by-products or residues that may accumulate in the stove. The main mechanism of regulation in this system is the rate at which fuel is added to the stove, if you regulate the combustion reaction by some other means, you will lose efficiency and produce by-products that are harmful to you and the stove.

Both primary and secondary combustion need to be managed slightly differently and take place in different parts of the stove.

All stoves designed for burning wood, should have two separate draught regulators: for primary and secondary air flow. The primary air is directed into the bed of the fire, at the char, and the secondary air is directed above the fire into the flow of volatile gases above it. As the volatile gases are released as the solid wood is heated, these tend to burn away before the char has finished burning.

LAYING THE STOVE AND LIGHTING UP

The ash bed

Before you start, it is best to leave a good bed of ash in the stove from the previous fire. This insulates the fire bed, and may capture small amounts of unburned material remaining from earlier use. This isn't always possible, as you will need to clean out the ash tray at some point, but it does make a noticeable difference to how quickly the stove will light (particularly in multi-fuel stoves which have very open grates).

Firelighters

The key to lighting a stove well is keeping everything dry. I don't have a lot of time for firelighters as you can get a satisfying fire going very easily with dry kindling and newspaper.* If you do use firelighters, I'd recommend the ones made of wood fibre and wax over the white oily ones. They burn much more cleanly and you don't get the nasty paraffin residue over your fingers.

Kindling

Kindling sticks are best kept indoors, because you want them to be as near to bone dry as you can get them. If you can get away with it, the airing cupboard is a great place to stash them (when your significant other isn't looking). Stacking them next to the stove on the previous night's burn also works well, or in a utility room with a dryer or boiler. Ideally split sticks are best for kindling as the split surface has a larger surface area than round sticks, and bark doesn't seem to catch quite as easily.

Laying the fire

Don't worry about leaving the fire laid. Unless you live north of the Arctic Circle, you won't ever be too cold to sort it out

* The ordinary stuff, not glossy magazines.

when you need it, and newspaper does seem to attract the damp when it's left under an open chimney.

Crumple the newspaper into balls and lay a bed of them first. Don't screw the paper up too small as you want enough air to get through to allow it to catch easily. You're probably looking for balls a bit bigger than fist sized. Make sure that there are a few edges sticking out, as these are the bits you'll want to light later.

Scatter kindling sticks over the newspaper. Make sure that they aren't stacked too closely, you want plenty of air gaps (like pick-up-sticks). I usually use a lattice arrangement. Don't be stingy with the kindling. I know plenty of people who want to get away without using any more than they have to, but for every time you get away with it, there's another when you end up with a heap of burnt kindling and a few slightly scorched logs. At this point you can either walk away for a bit, or try and start again while running the risk that the newspaper will spontaneously catch from the hot ashes while you're still laying the kindling...

After you've laid the kindling down, put a couple of small logs on top. If you have charred ends from a previous fire, they'll work well (though if you're burning dry wood, you're unlikely to have any). Otherwise add small, dry logs, about wrist thickness is good, preferably with at least one split face.

Lighting

First, make sure that all of the draughts are as far open as they will go. You want plenty of air to get in while the fire gets established. Then light the paper at the bottom of the fire in as many places as you can before the match burns your fingers.

They won't all catch properly, but you'll want at least two or three to start burning really well. Remember that the flames will want to travel up whatever they are burning, so light the bottom edges of the paper where possible.

As soon as your match has burnt out, shut the door. The draughts will begin to draw and suck air in. If you open the door at this stage (before you are generating enough heat) you're likely to get a face full of smoke, as the resistance to the smoke spilling into the room is much lower than the resistance to getting it up the chimney.

Occasionally you may find that the stove instructions indicate that you should leave the door open slightly at this point. This is to increase the available draught beyond the usual maximum primary and secondary air settings. It's always best to follow the manufacturer's instructions as they will be specific to the make and model of stove that you're operating. However; in this case, I would suggest being especially careful to use very dry kindling and paper to light the fire quickly (and generate a strong up draught as soon as possible).

When everything is burning well, open the door **slowly** (to minimise the amount of smoke and ash drawn into the room) and add more fuel. You are now in phase one burning.

Make sure you always keep the stove door closed, except when adding fuel, as the effective control of air into the stove is what gives you the high efficiency. An open stove is approximately equivalent to an open fire in terms of efficiency (i.e. not very good!)

PHASE ONE BURNING

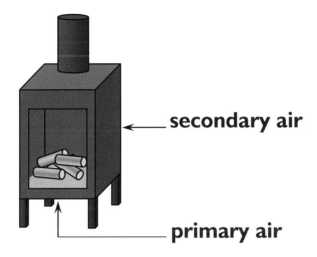

secondary air

primary air

Figure 10. Phase one burning

Phase one is an efficient fast burn. Volatile compounds and water are released from the wood as it is heated. Good secondary air and primary air supply is available to ensure that all released hydrocarbons are fully burnt and all flue gases are >120°C when they leave the flue. There should be little or no visible smoke, only a small amount of steam released. All of the water in the wood evaporates away during this phase.

PHASE TWO BURNING

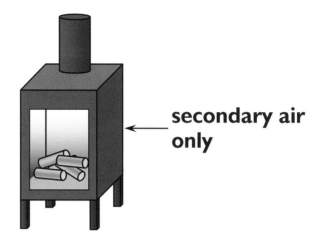

secondary air
only

Figure 11. Phase two burning

Phase two is an efficient char burn. This takes place when the volatiles and water have all been released. The overall air supply can be reduced at this point, but you should still keep adding secondary air. No primary air is needed. This allows the stove to continue to achieve complete combustion while cooling down. This state should be how you would usually leave a stove overnight. Stoves can stay in this state for a surprisingly long time, so don't immediately rush to add more logs as soon as you hit this point.

INCORRECT OPERATION

primary air only / no air

Figure 12. Incorrect stove operation

Wood stoves are commonly used poorly. Where fresh wood has been added with a limited primary air supply, volatile compounds and water are released but don't burn and remain cool (around 60°C). This generates a lot of tar deposits and smoke. This often occurs when people want a stove to stay lit overnight (see the section below for more details).

Another similar common problem is when the stove door is left open during operation. This has the effect of turning an efficient log stove, which is capable of good secondary combustion, into an open fire, which loses a significant proportion of the fuel's calorific value up the chimney. This is another cause of incomplete combustion and will reduce the life expectancy of your stove.

KEEPING STOVES 'IN' OVERNIGHT

I often come across people who want to know how they can keep a woodburning stove producing heat overnight. The short (and unpopular) answer is that either you'll have to keep adding fuel, or you should have a stove installed that is designed to retain heat for a long time. Other methods involve reducing the efficiency of the reaction in some way.

Here are the methods I have come across for keeping a stove producing heat overnight. Some of them seem pretty sensible, others less so.

1. Keep adding fuel at regular intervals overnight. Not (quite) as ridiculous as it sounds, this method will keep a good heat output going and the stove will stay efficient and producing the maximum amount of heat possible. I have used this method when I was staying in a friend's tent that came complete with a Norwegian army surplus stove. The weather outside was cold and wet and it was by far the best way to stay warm. At three in the morning, the sight of the stove glowing red was very comforting.

2. Get a stove that is designed to retain heat. There are a number of traditional stove designs that are designed to incorporate a large amount of thermal mass, this retains the heat from the stove long after the fire has died down. They are usually quite large and usually constructed of masonry or tiles (a good example would be Northern European Kachelofens). These appliances work as mass storage heaters which release heat slowly, heating a much larger space than conventional stoves, and are often designed to heat whole homes as opposed to a single room. It's important to note that while these stoves do retain heat

wonderfully, the quantity of thermal mass involved means that they can take a very long time to heat up. In some cases this can be continual firing for many hours before the benefits become apparent. While this is extremely useful in continental climates, where it is common to need a heat source for twenty four hours of the day, UK heat use tends to be much more inconsistent due to a changeable (and relatively mild) climate. This can lead to difficulties of having too much heat when it is not needed, or not being able to produce heat fast enough when it is. If you are comfortable with using other forms of storage heater, then these may be an option worth investigating.

It is important not to confuse these traditional stove designs with newer stove models which use similar principles. A number of new stoves are coming onto the market, make use of modern materials to eliminate a number of issues that are commonly associated with traditional designs (such as the length of time required to get to temperature and cracks appearing). Using materials such as silicon carbide (which is very hard and has a very high resistance to corrosion) these appliances can produce highly efficient modern stoves which retain heat more effectively than conventional designs.

3. Add green logs to the fire. In my opinion, this is a bad idea. Green logs dampen down a fire by increasing the amount of water that needs to evaporate before they can burn. While this does increase the length of time it takes for them to be completely consumed, it also reduces the temperature of the fire and increases the proportion of steam in the flue gas. This combination of factors is likely to cause incomplete combustion and increase the particulate matter and carbon monoxide given off; as well

as causing an increase the level of condensation in the flue. Wet products of combustion form tars and residues which are acidic and will shorten the life of your system. I've seen double wall stainless-steel flues that have completely corroded through in as little as five years from this treatment.

4. Add something to the fire that naturally burns longer and retains heat better than wood. This is usually coal. There are number of good reasons for not burning coal (particularly on a stove that is only designed to burn wood), but the fire may (just) stay in overnight.

5. Restrict airflow. Restricting the amount of oxygen reaching the fire will slow the reaction and allow the fuel to last longer before the fire dies. Again, in my opinion, this is not a good idea. In restricting the airflow, you are placing the fuel in a low oxygen environment while maintaining a high temperature. These are ideal conditions for incomplete combustion (this is effectively the method used to make charcoal). Cooking the fuel rather than letting it burn will result in the release of harmful flue gases like carbon monoxide.

6. Accept that the stove will go out and concentrate on retaining the heat generated when it was still lit. This can be very good advice, particularly if you have a poorly insulated, draughty property. The most cost-effective method of keeping a house warm is usually to insulate first, before upgrading any boilers stoves or other heaters.

5. Maintenance

ASH REMOVAL AND DISPOSAL

If you are only burning clean wood, then ash disposal is not a problem. Ash from clean wood is a low nitrogen fertiliser and is quite safe to put on your compost heap (though do try not to breathe it in). If you find that the ash blows all over the place when you're removing it from the stove, I'd recommend investing in a plant mister to **very slightly** dampen the top layer and prevent it from flying around. Be careful not to get the firebrick lining of the stove wet as this will tend to make it crack when you light the stove. **Don't** use your vacuum cleaner to empty out the fireplace unless you are absolutely sure it is cold. Spark resistant cleaners are available, but they are very expensive and not really necessary for a domestic setting.

Coal ash, and the ash of anything that isn't virgin wood (i.e. as it grows in the tree, with nothing added) is suspect, and a lot less nice to have lying around. It is much more likely to have unsavoury components such as heavy metals in it and should really go straight to landfill. Again, do make sure it's completely cold before you put it in the bin. I suggest you leave it in a metal bucket outside the back door for a couple of days to really make sure (and if it gets rained on, so much the better).

A note on uses for ash

Clean wood ash has a number of uses, these vary from simply where you dispose of it, to some that need you to be a bit more keen to try. As a chemical it is alkaline and caustic, it is also an abrasive, which makes it useful for a range of household jobs.

■ Fertiliser

The most common use for wood ash is as a low-nitrogen fertiliser (potash). This means that at the very least, you can dump it in the corner of your garden with the grass clippings without worrying that it's going to cause a problem.

Ash is commonly added to compost, this can be a bit of a mixed blessing for two reasons:

Many of the chemical constituents of wood ash are highly soluble in water, so if you have a compost heap that is uncovered, you may find that the increase in fertility in the compost is negligible, while the surrounding area becomes a bit of a jungle.

Because wood ash is alkaline in nature, it will tend to react with acidic compounds within the compost heap (these tend to crop up in materials which are high in nitrogen such as fresh manure, and grass clippings). This reaction releases ammonia that not only removes nitrogen from the compost, it also smells pretty awful.

Instead of adding ash to the compost, you may find that it is better to add it directly to the soil, where any leaching will result in a local increase in fertility as the soluble compounds will be immobilised by uptake into growing plants. It's probably worth mentioning that if you are on particularly alkaline soil, you should be quite careful about adding ash to the soil, as you may find that you are actually decreasing fertility, unless you add an acidifying agent as well.

■ Pest Control

Wood ash is an organic slug-deterrent, leaving a ring of wood ash around (but not touching) vulnerable plants is reputed to keep the slugs off. This may because of the chemical composition that will tend to dry slugs out (the same as salt or vinegar) in which case, repeated applications would be necessary after rain (and you'd have to be careful that you're not adding too much alkaline material to the soil). Another potential cause would be because the abrasive nature of the ash is uncomfortable for the slugs to cross. I have to admit I have my doubts about this as a plausible effect as slugs seem to be quite capable of bypassing this sort of barrier, and I've seen ground eggshells (a similar method) fail quite spectacularly.

The soluble alkalinity of wood ash also makes it an anti-algal agent for use in garden ponds. I have no direct experience of using this, but I'm told that a little goes a long way. I would suggest doing some further research before you start adding it by the bucket load.

■ Abrasive

The silicate component of wood ash is a mild abrasive. This can be used for a variety of household cleaning jobs as well as being the traditional method of keeping stove glass clean. It has also been used to treat paths in winter where its soluble component helps to melt ice as well as providing a mechanical grit.

■ *Producing lye (potassium hydroxide)*

Lye is a traditional ingredient in soap making. It is made by dissolving the soluble components of wood ash and then concentrating the resulting solution (usually by boiling). This is then reacted with some form of fat (traditionally animal suet) to form a soap.

SWEEPING THE CHIMNEY

According to most chimney sweeps you should have your chimney swept at least once every six months. If you use the stove a lot, you may need to have it swept more often. Failing to sort this out can lead to soot build up in the chimney (especially if you're burning inefficiently) that, in turn, leads to corrosion and increases the risk of a chimney fire. If you're feeling especially keen, you can go and get a set of brushes and do the whole thing yourself, but to be honest, I think I'd rather pay to have the expert who knows how to do it cleanly without getting soot all over my living room. It's just another one of those jobs, like cleaning the gutters out, that nobody really likes, but everybody has to do one way or another. The three national organisations representing chimney sweeps are:

- The National Association of Chimney Sweeps (www.nacs.org.uk).

- The Guild of Master Chimney Sweeps (www.guildofmasterchimneysweeps.co.uk).

- The Association of Professional and Independent Chimney Sweeps (mainly in the south west, www.apics.org.uk).

All three organisations have lists of members online to help you track down a sweep local to you.

FIRE ROPE

Fire rope is the soft, heat-resistant woven cord that forms a seal around the stove door. It does have a shorter life expectancy than the main body of the stove, and you will need to replace it from time to time. You can often get this done by your chimney sweep as part of a servicing package. Failing that, many manufacturers supply parts online. This is quite cheap; you can probably expect to pay something in the region of £10 + labour, at the time of writing.

FIREBRICKS

The ceramic lining of your stove is made of firebrick. This helps the stove to retain heat, for longer than it would other-wise, and provides some resistance to corrosion. Firebrick is very durable in terms of temperature, but is quite fragile mechanically. They can often be cracked and broken by logs being added to a fire a bit too enthusiastically, logs shifting as the fire burns through, or through being heated if they have got wet (from poor quality fuel).

Many manufacturers offer replacement firebrick sets that are pre-cut to the right size. If you have a non-standard stove, you should be able to get a custom set fitted by a stove installer. The price is pretty variable depending on the stove design, and complexity of the shapes required. Some manufacturers only sell complete sets as well, which will tend to push the cost up.

CLEANING THE GLASS

The main advantage of stoves without a glass front is that you don't have to do this job. It is more or less inevitable that any glass on your stove will become covered in soot and need cleaning (in spite of any clever air-wash design). There does seem to be a fairly healthy debate about the best methods for cleaning, so there is a range of options to try. The methods I've come across are:

Damp newspaper and ash

By far the cheapest option, in my experience this works pretty well for the looser material, but doesn't have much effect on the heavier stains. There are some variations to this method, such as removing the glass and soaking it (or covering it in wet newspaper) overnight, which should help a bit.

Vinegar

I've heard a few people swear by this as a variation to the newspaper and ash method. Vinegar is fairly commonly used as a de-greasing agent to clean windows, so I don't see any reason why this shouldn't help. Alternatively, I've heard of people using lemon juice, it's not clear what advantage that this would have over vinegar except that it will leave your stove lemon-fresh...

Cleaning products

There are a number of products on the market that you can use. Remember that you want the least aggressive product that will do the job. There are plenty of abrasive scouring powders and cream cleaners available, but many of these will make a bit of a mess of the glass as well. The best option I've come across is oven cleaner that's designed to spray on and wipe off without scrubbing, although you probably do need to remove the glass before using it, which turns the whole process into a bit of an undertaking.

Mechanical scrubbing

Sooner or later, most cleaning jobs come down to elbow-grease. The critical thing to remember here is that if you abrade or scratch the surface of the glass, you're going to have a devil of a job cleaning it next time. I'd suggest plastic pan scourers, though you still need to be careful (don't forget that the soot itself is abrasive). Avoid wire wool or any metal scrubbing brushes like the plague.

6. Fuel Management

HOW MUCH DO I NEED?

I get asked this question a lot. Unfortunately it's a bit like asking 'how long is a piece of string?'

There are a number of conversions you can do to give you an idea of the amount of wood you have, and the amount of heat that it represents. Bear in mind though that this will only give you a rough estimate: fuel quality is variable, as is the weather and your heat demand. With experience, you'll be able to estimate how much fuel your stove, and average usage, requires.

There is an extended version in Appendix 1, with all of the formulae and calculations. Here is the short version.

Table 4: Calorific values of wood by weight and volume

	green	part-seasoned	fully seasoned	kiln-dry	Wood Briquettes*
Calorific Value (kWh/tonne)	2,300	3,190 – 3,490	3,790 – 4,090	4,680	4,680
Calorific Value (kWh/solid m³)	2,210	2,360 – 2,390	2,420 – 2,450	2,500	5,150

The volume measurement above is based on an average oven-dry timber density of 480kg/m³ and a solid cube of wood (1m by 1m by 1m with no air gaps). In practice loose logs occupy about 2.5 times the space that a solid block would, and neatly stacked logs take up about 1.5 times the space.[†]

To work out how much energy is in a quantity of logs, you would need to take the standard calorific value for a solid m³ of timber and divide it by either 2.5 or 1.5 depending on whether the logs are stacked neatly or loosely heaped.

So, 1m³ of green logs contains:

2,210 kWh of energy (if it is a solid lump, in which case you have some serious sawing to do),
or
$2,210 \div 2.5 = 884$ kWh of energy if it's heaped loosely,
or
$2,210 \div 1.5 = 1,473$ kWh of energy if it's stacked neatly.

Remember these figures don't take into account the efficiency of your stove (likely to be around 70 per cent if properly set up).

What does that mean per year?

The exact figures will vary depending on your situation, but I've made some assumptions and done a rough calculation here to give you an idea of the method.

■ *Step 1. Determine usage*

Assuming daily usage during a heating season that lasts from the 15th of October to the end of February for an average of three hours per day (or 408 hours of usage/winter) at 70 per cent efficiency.

---

* Approximate values. The true value will vary depending on the feedstock used to produce the briquette (and in the case of volume, the diameter and shape of the briquette).

† Except for briquettes which, because of their regular size, will usually pack 2.0 for loosely heaped and 1.2 for neatly stacked.

For a 3 kW-rated stove, this equates to 408 x 3 = 1,224 kWh of heat produced annually. The same usage pattern for a 5 kW stove would mean 408 x 5 = 2,040 kWh produced annually.

- ## Step 2. What quantity of logs is required?

While both stoves may be producing 3 and 5 kW respectively (a measured amount) we know that they are probably only running at around 70 per cent efficiency. This is going to increase the fuel requirement proportionally:

([heat output] ÷ [calorific value per unit]) ÷ [efficiency]
= number of units of fuel required

So for a 5 kW stove, producing 2,040 kWh of useful heat per winter with a fuel calorific value of 4,090 kWh per tonne, the calculation would be:

(2,040 ÷ 4,090) ÷ 0.7 = 0.43 tonnes of fuel

I have done the calculations based on a range of values for both a 3 kW and a 5 kW stove below:

Results By weight

Table 6. Example fuel requirement calculation by weight

	Tonnes of fuel required
3 kW stove	0.21 – 0.23
5 kW stove	0.50 – 0.54

- ## Step 3. Calculating volume

The volume calculation requires an extra step.

Fully seasoned wood contains between 2,420 and 2,450 kWh per solid m³.

Table 7: Example fuel requirement calculation by solid m³

	Solid m³ of fuel required
3 kW stove	0.50 – 0.51
5 kW stove	0.83 – 0.84

To translate this into a load of logs, we need to multiply the results by the stacking densities above (1.5 for stacked, 2.5 for loose).

Table 8: Example fuel requirement calculation by stacked/loose m³

	Neatly stacked (m³)	Loosely heaped (m³)
3 kW stove	0.75 – 0.76	1.25 – 1.26
5 kW stove	1.25 – 1.26	2.08 – 2.11

Of course if you have better figures for your heat demand, and fuel moisture content, you'll get a more accurate estimate.

STORING FUEL

When you have bought a load of logs, the last thing you want is for them to deteriorate and lose calorific value before you use them. This is even more irritating if you've sweated to cross cut, split and stack logs that you've produced yourself.

Outside stores

The main requirement for log storage is a well-ventilated dry space. This doesn't need to be inside, although a shed or garage would be perfect. Logs will need to be protected from rain, and from water drawn up from the ground. Snow is a particular problem because while it melts it holds water in contact with the log surface and allows it to soak in.

The easiest way to protect logs from picking up water from the ground is to stack them on old pallets, but any old long, straight(ish) pieces of wood will make adequate bearers, this can provide a use for any old fence posts you have lying around (that can't be burnt because of preservatives). Stacking on a tarpaulin is not a great idea because, in my experience, it tends to collect water – leaving your dry logs sitting in a puddle.

As well as protecting your logs from groundwater, you'll also need to cover them with something like an old bit of tarpaulin to keep the rain (and snow) off. It's useful to try and keep the tarpaulin angled to shed the water instead of allowing it to pool on the surface. You don't need to bother covering the sides of the stack, as letting the wind get through will help the drying process, but if you can orient the stack to allow one of the longer sides to face the prevailing wind/sun direction, this will help.

Remember that you need to keep the stack ventilated. Sealed containers, such as old wheelie bins, are great for keeping the rain off but don't let any water in the wood escape. If you don't have any ventilation the logs tend to sweat condensation onto the inside of the container, which then drips off, rewetting the logs and undoing all your good work.

- *Location*

The primary consideration for the location of your log stack should be convenience. There seems to be some kind of natural law that dictates that your indoor log supply will run out at the exact time the rain really starts coming down, or as soon as it's too dark to work without a torch. Having said this, if you are able to stack your logs in a space that is less likely to collect rain or snow, or lose the tarpaulin in strong winds, then so much the better.

The traditional method of storing logs is often to stack them against a convenient wall. This helps to keep the rain off, and also helps to insulate the house. If your house does have poor insulation, then you can take a small crumb of comfort knowing that some of your lost heat is probably helping to dry out your log supply. You still need make sure that the pile is covered and off the ground. I once visited a gentleman (who shall remain nameless) who had carefully stacked all his logs in a lovely neat pile against the wall of his house, only to realise that his gutters were leaking all over it. It's also worth keeping the stack away from the wall slightly to prevent damp in the brickwork. It's less of a problem in newer properties, but it can happen, particularly if you only have single-skin brickwork.

Inside Stores

From a fuel quality perspective, storing fuel inside is ideal. If you can dedicate a space in a garage or shed, then these can become a really useful space for fuel storage. I may be in a minority, but I'd far rather keep my fuel dry, than worry about my car getting wet (cars are after all designed to be waterproof.) The key elements to get right in these spaces are just the same as for outside stores, prevent the stack from re-wetting and make sure that the ventilation is good enough.

Re-wetting is not really an issue indoors, but it is still probably worth keeping the stack on bearers to ensure good ventilation through the fuel. Ventilation shouldn't be too much of a difficulty in the average shed or garage, as they are often quite draughty spaces, but if you have a well sealed area like a porch, you might find that you are getting a lot of condensation. This is not really an issue for the fuel, but it is could cause trouble with the fabric of the building. Remember that green wood is about 50 per cent water – so for every kg of green wood you are bringing into the house, you're bringing half a litre (or nearly a pint) of water in with you.

Heated areas

The same rules apply for storing logs in a heated space, and it's quite uncommon to try drying significant amounts of wood in a heated area. The exceptions to this are a small store of dry logs which are due to be burned soon, and kindling. Both of these cases relate to fuel that has already been dried, so the risk of causing significant damp problems is minimal.

Do take care when stacking wood next to a stove that is running. Some appliances can get hot enough to cause wood or other materials to ignite when in contact with the stove's surface. This is not a very common problem, but some appliances (particularly older or cheaper models) are vulnerable. If your stove does get this hot, I would suggest a fireguard as a matter of course to prevent anything (or anyone) accidentally brushing against the surface.

Storing briquettes

Briquettes are a more refined type of fuel and will need you to take a bit more care when storing them (you should never store briquettes outside):

■ *Moisture*

Briquettes have been dried to an artificially low moisture content and are highly absorbent. They almost always come in airtight plastic packaging, and you should keep them in the packaging until you need them. If briquettes get wet, they not only absorb moisture, but the wood fibres also begin to swell. This causes the briquettes to crumble and turn back to sawdust, effectively rendering them worthless as fuel. Do examine the packaging when they are delivered and use any that have been damaged first (as damaged packaging will mean that they are more vulnerable to moisture when stored).

■ *Mechanical damage*

Briquettes are far more susceptible to mechanical damage than logs. Again, any rough handling is likely to turn them back to sawdust. The amount of sawdust present in the bottom of the bag when you receive briquettes is a fairly good indication of quality, as poor briquettes will either not hold together as well in the first place, or may have been roughly handled in transit.

7. Price

The easiest way to compare price between different sources of timber is in pence per kilowatt-hour (p/kWh). This takes into account the variation in the calorific value between loads due to moisture content or density. The other handy aspect of this calculation is that it allows you to compare the price of heating using wood with that of using other fuels. Gas and electricity are both sold in kWh and heating oil has a very easy calculation (1 litre of oil = 10 kWh).

There is a detailed guide to this calculation in Appendix 1, or you can find details on the BEC website (see Appendix 3 for more details.)

Effectively this is an exercise in determining the number of kWh you are buying and using that figure, and the price you're being asked to pay, to match like with like. At the time of writing, domestic fuels only attract a VAT rate of 5 per cent so keep an eye on what is quoted on your bill.

CONVERSION BY WEIGHT

For weight it's just a simple matter of multiplying the calorific value per tonne by the number of tonnes and then dividing by the asking price:

$$\frac{CV_t * N_t}{(£ * 100)} = p/kWh$$

CV_t Calorific Value/tonne
N_t Number of tonnes in a delivery
£   Asking price in pounds (including delivery cost)
p/kWh Pence per kWh figure

CONVERSION BY VOLUME

For volume, it's a little more complicated because you need to add an additional element to the calculation: the packing density (how tightly the logs are stacked) as described earlier.

$$\frac{(CV_m/d) * N_m}{(£ * 100)} = p/kWh$$

CV_m Calorific Value / m^3
d   packing density (2.5 for loose and 1.5 for neatly stacked)
N_m Number of m^3
£   Asking price in pounds
p/kWh Pence per kWh figure

8. Quality fuel characteristics

MOISTURE CONTENT

This is really, really, important. Bear in mind that there are a lot of mistaken ideas (and downright lies) about the moisture content of logs. Even if you're sure the logs you're getting are what has been advertised, 'seasoned for a year' just means that the tree has been felled a year ago – it doesn't give any guarantee of where or how it's been stored, etc.

Essentially, you're looking for the driest logs available at the right price.

Moisture meters

Moisture meters are a bit of a mixed bag. As far as I've been able to tell, you can get good results from one costing £15 as easily as one that cost ten times that. Unfortunately, you can also get bad results just as easily. At best, moisture meters offer an indication of the moisture content of the piece of wood you test (not the whole load). There are a few things you'll need to know before you use one.

- *What basis measurement does it show?*

There are two different ways of calculating a moisture content percentage, 'wet basis' and 'dry basis' (the difference is explained in Appendix 1). Make sure that your moisture meter is testing a **wet basis** measurement. The Appendix has information about how to do this.

■ *Calibration*

Ideally, you should check your moisture meter against a piece of wood with a known moisture content to check it's reading correctly. The Biomass Energy Centre has a useful simple guide to testing moisture content available for free on their website.*

■ *Taking a measurement in the right place*

If logs have been recently cut down from longer lengths, the moisture content in the middle of the length will be much higher than at the ends. You need to make sure that you've got a good representative log from the stack. This is an occasion where more is better. As long as your samples are representative of the load, more data will mean a more accurate result.

Ideally you will want to split a few logs and check the inside face. This is to make sure you haven't got really wet logs that have been standing in the sun for a few days, or really dry ones that got rained on in the lorry.

* www.biomassenergycentre.org.uk/pls/portal/url/ITEM/
B8635B4D13090FAEE04014AC08042CC8

CONTAMINATION

Fuel with painted material or chipboard or tanalised timber is bad news. A few stones or a bit of soil is probably OK, though if you're getting lots of them it's going to be a problem. In particular, watch out for stones when you're cleaning out the ash, they hold the heat much longer than their surroundings and can give you a nasty burn. Essentially, anything that isn't clean wood is best avoided.

LOG SIZE

This largely depends on the size of the stove you want to burn them in. Anything too bulky is going to end up being trouble either because it won't fit properly, or because it will reduce the surface area of the fuel in contact with the fire. Big logs are also a bit more likely to be damp as they'll take longer to dry than smaller ones.

SPECIES

Personally, I really don't have a problem with most species, particularly if they are dry and in an enclosed firebox to contain spitting. Pretty much any wood will give you a satisfying fire if it's been properly dried and cut. If I was forced to choose, I would probably either go for the densest wood I could find, usually oak, or if the price was too high (oak is often in demand) probably something easy to get hold of, like larch.

PESTS AND DISEASES

The threat of pests and diseases in British trees is very prominent at the moment. A number of new pathogens have become established in the UK and are likely to cause significant damage to woodlands. Understandably, this has raised concerns about transport of all timber products within the country as potential routes for infection. The forestry industry is acutely aware of the dangers of these pests and diseases, and there is a wide range of measures already in place to reduce their spread or minimise the risk of further infections. There is a list of the most significant threats on the forestry commission website here: www.forestry.gov.uk/website/forestry.nsf/byunique/infd-6abl5v and guidance on biosecurity here: www.forestry.gov.uk/biosecurity

If you're worried that trees near to you have been infected with a disease, you can report them directly to the Forestry Commission here: www.forestry.gov.uk/treealert

There are also a few useful hints to minimise the risk of handing material:

Green wood

The majority of tree pests and diseases occur either as a pathogen in a living tree or in the decomposition of rotting wood. In both cases, a relatively high moisture content is required to allow these organisms to flourish. The most straightforward method of preventing decomposition is drying the wood as thoroughly, and as quickly, as possible. This also has the benefit of preserving the greatest possible calorific value as, at its heart, any decomposition is the conversion of the chemical energy in the wood into the metabolic process of the organism responsible.

To be honest, burning rotten wood is pretty poor practice anyway, as it is often very wet and difficult to burn. If the wood isn't sound, you're best to leave it alone.

Don't burn in an open fire

I have already listed a number of good reasons not to use open fires on page 10 above. In addition to these, using a device that does not ensure complete combustion poses a (small) risk of distributing fungal spores over a wide area. This is because the heat generated may be insufficient to kill spores, while the convection currents generated by the chimney provide a ready-made method for diffusing particulate matter over the surrounding neighbourhood.

Don't collect leaf litter

Leaf litter is often recommended by gardening sources as a good additive to soil, either as a mulch or as an ingredient of compost. This is because adding cellulosic (woody) material to the soil improves its structure, and because many plants make use of a symbiotic relationship with *mychorrizae** that leave spores in leaf litter. Unfortunately, there are a number of pathogens which can be transmitted in exactly the same way, ash die-back (*Chalara fraxinea*) being the notorious example. I would certainly be **very** cautious (verging on the paranoid) about collecting any leaf litter from sites outside my garden.

---

* Fungi which act as an extended root network in return for access to the products of photosynthesis.

BRIQUETTE VARIABLES

There are a few variables that are only really an issue for briquettes:

Species

Rather than worrying about which type of wood you're getting, when you buy a briquette the worry is often that you're not getting wood at all. While non-woody species are often much cheaper than wood-briquettes, the price is usually that you are reducing your stove's life expectancy (and may be invalidating the warranty). This may not be an issue if you are prepared to accept the risk, but it is something you should be aware of.[†]

Packaging

Briquettes almost always come in some form of airtight packaging (usually shrink-wrapped plastic), as they are highly absorbent and will tend to gain moisture if left in ambient humidity. This additional packaging will increase the amount of waste that goes to landfill.

Mechanical durability

This is a big deal, poor quality briquettes will tend to break more easily and crumble down to sawdust (which basically makes them worthless as fuel). Crumbly briquettes may also be an indication that they have got wet at some point. A small

---

[†] If your stove is not suitable for burning non-woody species, you may also be breaking the law if you are inside a smoke control area.

75

amount of sawdust in the bottom of the packaging is normal, but if it is excessive, it's usually a good indicator of a poor quality fuel.

Energy Balance

Briquettes do take more energy to form than logs and this does raise concerns about the long-term sustainability of production. It's perhaps not as bad as you might think, however, as the material being used to produce briquettes is often a by-product of some other timber processing business. It's relatively safe to say that the sawdust used in manufacture would be produced in any case, and would be sent to landfill if not otherwise used. The comparison therefore is between the energy cost of turning sawdust into a briquette, packaging and shipping (while displacing some other fuel use) as opposed to transport to landfill and decomposition while retaining the other fuel.*

* Do look at the facts and figures section of the Biomass Energy Centre website if you want some more details. www.biomassenergycentre.org.uk

9. Buying fuel

There are two main types of buyer in the log market:

1. Buyers who see logs as a primary source of heat for their houses. These people will tend to buy as cheaply as possible in bulk. The price of logs brought by these people needs to compete favourably with the price of oil and gas.

2. Buyers who have a stove or open fire for amenity use. This group won't be using wood as a primary heating source and may be only using the stove very occasionally. This means that they are less likely to have a good grasp of the current price of logs, and are likely to buy in smaller loads. Since this amenity use is more a treat than a necessity, the price of logs that fall into this category tend to be competing with other treat activities such as going out for a meal or going to the cinema (i.e. they are more expensive). Most kiln-dried logs and logs sold in petrol stations fall into this category.

Ideally, if you're going to be using your stove a lot, you'll want to find a supplier that can deliver a good quantity at a reasonable price, and if you have space, set up your own system for seasoning the wood as thoroughly as possible (rather than paying extra for premium quality fuel, or small quantities at inflated prices).

WHAT'S IN A LOAD?

If you are fortunate enough to have a vehicle or trailer that can handle the weight, you may find that some suppliers are prepared to give you a discount if you save them the time and trouble of delivering. It also means that you'll have a good idea of how much you're getting. Log suppliers are notorious for supplying logs in an imprecise unit: the 'load'. This usually translates into an approximation of what they can fit in the trailer (loose). It's always a good idea to check how big the load is before ordering.

Some of the premium log merchants have taken to delivering logs pre-stacked in a container so you get a batch of logs complete with their own storage container. While this can be very convenient – no stacking or building log stores – you are certainly going to be paying a premium for this service.

KILN DRIED OR NOT?

There are a number of suppliers who offer kiln-dry logs. These are premium logs that have been artificially dried in a wood kiln to a lower moisture content than is possible using the air-drying method. This is not automatically as bad for the environment as you might expect as this category can also include waste off-cuts (which would otherwise go to landfill.) A number of suppliers of this sort of logs also use biomass to heat the kiln; though whether this is better for the environment is still debatable as at best this is adding energy to a process when it is not necessary (20 per cent air-dried is good for most purposes). Kiln-dried wood is also often imported from the Baltic states by some of the bigger suppliers (with the associated transport emissions). As such, it is usually safe to assume that you will be paying a premium for kiln-dried wood, you may also want to check the sustainability of this material more thoroughly.

BUYING GREEN?

Green logs should be cheaper than dried logs as you're not paying for someone to store them and dry them for you. Do bear in mind that the logs will lose a good deal of weight during the drying process though, so the figures may be a little more complicated than you'd first expect. (There is a formula for calculating how much in Appendix 1.) Tree surgeons are a good source of green logs, if you can persuade them to keep them whole rather than just chipping the whole thing. I've included a section on drying your own logs later.

WASTE WOOD

Waste wood is a bit of a mixed bag. If you don't care what the wood in your stove looks like, it can be a fantastic, cheap way of getting kiln dried firewood. The main thing to watch out for is contamination. Nails, grit, or stones are nothing to worry about, but the last thing you want is a bunch of wood that you have to throw away because it's full of paint, preservative anti-fungal agents or other chemical contaminants.

Where to find waste wood:

• Joinery shops

• Sawmills

- Companies dealing with large volumes of pallets

- Furniture, door and window frame manufacturers

- Anyone making timber roof trusses.

The advantage of many of these sources is that they can be happy for you to take away something that for them is a headache (i.e. waste disposal) and potentially even give them something for it in return. Bear in mind though that a lot of people producing wood waste may have already started marketing it to fuel suppliers or be using it themselves – so you might have to look a bit harder than you first expect. Still, it does no harm to ask and I've had enough people on the phone wanting to get rid of wood waste to know that there is still a lot of it floating around.

BRIQUETTES

Briquettes can be one of the most expensive ways of fuelling a log fire. That said, if you need something dry with a good calorific value to offset poorer logs they can be very useful. The problem is often finding out an accurate figure for the calorific value, particularly if the supplier is blending together different species (e.g. wood and *miscanthus*). In my experience, wood briquettes seem to be quite expensive and the cheaper versions come from a range of non-wood products.

FINDING A SUPPLIER

Supplier accreditation

There are two types of accreditation: forestry certification and fuel accreditation. Forestry certification is an independent audit of forestry practices against sustainability criteria. An example of this is the FSC Certification Scheme.* While this form of certification gives an assurance of sustainable management, it is expensive to undertake, which means that on the whole, only owners of larger blocks of woodland or management companies apply. In addition to this optional accreditation, The UK Forestry Standard (which all UK forestry operations have to follow) is based on internationally agreed 'sustainability criteria', this has led to the FSC classifying UK woodland as 'low risk'.

The absence of certification is not the same thing as evidence of unsustainable production. You should also bear in mind that this form of certification only applies to the woodland management and not ongoing processing, so there is no guarantee that the fuel will be any good when you get it home.

Fuel accreditation is a different animal altogether. Under this system the product is being assessed for quality, so you know that you're getting a dry log, if that's what you've paid for.

Since the first edition of *The Log Book*, there have been a few changes to supplier accreditation in the UK. The two schemes that were formerly in place have joined forces and begun collaborating under the 'Woodsure' brand (www.woodsure. co.uk). This scheme now covers all forms of woodfuel,

* www.fsc-uk.org

including logs and briquettes.*

Woodsure operates with HETAS to assess fuel quality and the supplier's competence, as well as checking the legality and sustainability of fuel production. It's important to note that this is not as rigorous as a full forest certification, as it is based around the felling licence system rather than a full woodland management audit, but it should give you a reliable indication that the fuel you're using comes from a professionally managed source. There is a list of Woodsure members on their website.

Directories

- The Biomass Energy Centre has an extensive list of suppliers of all types of woodfuel at www.woodfueldirectory.org

There a number of other online directories that can help you find a supplier of logs, though some of them only supply information in specific geographical areas and may charge businesses for entry (limiting the number of details they supply):

- The LogPile website
  www.nef.org.uk/logpile/fuelsuppliers/index.htm

- South West Woodshed
  www.southwestwoodshed.co.uk/index.php/search

- Stoves Online
  www.stovesonline.co.uk/services/firewood-suppliers.html

---

* Though only briquettes made from wood. Straw and *miscanthus* briquettes are **not** covered.

- West Sussex log suppliers directory
  www.westsussex.gov.uk/leisure/explore_west_sussex/
  wildlife_and_landscape/trees_woodlands_and_
  hedgerows/woodland_products_-_west_susse.aspx

- Woodfuel Wales
  www.woodfuelwales.org.uk

Buying from the Forestry Commission

The Forestry Commission doesn't at present sell firewood logs. There are a number of forest districts that allow members of the public to pick up fallen wood for a nominal fee (scavenging permits), but this material will certainly need cross cutting and drying before it can be used. I've included a bit more about this in the section on buying raw material below.

Forestry contractors and tree surgeons

There are a couple of trade organisations representing forestry contractors and tree surgeons:

- Forestry Contractors Association
  www.fcauk.com

- Arboricultural Association
  www.trees.org.uk

Failing that, word of mouth works quite well with tree surgeons, or you could try the local paper or Yellow Pages.

The local pub test

I have a friend who swears by this method (although I do suspect that he might have ulterior motives). His theory is that pubs don't have time to be messing around with wet or poor quality logs, so the easiest way to find a good log supplier is to go and ask at the nearest pub that has an open fire. Of course, it is necessary to have a good look at the fire for a while to check that the logs are burning well, and it would only be polite to buy a pint while you're waiting... His thinly veiled excuses to go to the pub aside, there may actually be some truth to this. I suspect, though, that while this is a great way to find a supplier of dry logs, it doesn't take into account what the log supplier is charging (or the price of any beer consumed).

The petrol station

I have to admit that I hate, loathe and detest buying logs from petrol stations. I have always found the price far too high and I rarely get any good fuel to speak of.

To put things in perspective, paying £5 for a netting bag of logs is the equivalent of paying around £400/m³. When you couple to that the poor chance of finding well-seasoned wood, you are likely to be getting off to a bad start. While it may be the case that you can get well-seasoned logs at a good price from a petrol station: I have never seen it.

10. Producing your own fuel

GROWING YOUR OWN

A conservative estimate of average woodland production is around 8m³ per hectare per year (m³/ha/yr) for coniferous tree species, and about 4m³/ha/yr for broadleaves. The actual figure will vary throughout the life of the tree: growth in year one will be nowhere near as fast as this, while growth in year thirty may be much higher. This mean volume is (relatively) constant between different species, but bear in mind though, that the weight will change significantly depending on moisture content. You should also remember that most forestry operations will not be looking to generate fuel alone, and straight, clear timber should normally fetch a higher price when sent to alternative uses. There is a document on the BEC website that you might find helpful. It gives information on the moisture contents and calorific values of a wide range of different UK forestry species.[*]

Growth rates are closely related to species, site conditions, and management techniques. The Biomass Energy Centre has a handy simple guide to forestry on its website.[†]

The Sylva foundation have a useful online tool to support good forest management called MyForest: www.sylva.org.uk/myforest

[*] www.biomassenergycentre.org.uk/pls/portal/url/ITEM/ B8635B4D13170FAEE04014AC08042CC8

[†] www.biomassenergycentre.org.uk/portal/page?_pageid=75,409221&_dad=portal&_schema=PORTAL

The Small Woods Association provides advice and support to owners and managers of small woodlands: www.smallwoods.org.uk

BUYING THE RAW MATERIALS

Forestry Commission

The Forestry Commission has a somewhat piecemeal app-roach to firewood sales. This varies by country (Scotland, for example, seem to be quite happy to sell scavenging permits for firewood) and by region. In some cases, there are ancient rights for people living within particular areas (The New Forest and Forest of Dean, for example). Rather than give you incomplete or inaccurate information I'd suggest you get in touch with your local office. There is a list of contact details here: www.forestry.gov.uk/forestry/HCOU-4U4HZV

You are able to buy trees from the Forestry Commission as part of their standard timber sales, but realistically this is for businesses only, because the scale of operations is far too big for a single log stove.

Other sources

• Local authorities and national parks may be able to sell you small volumes: try asking your local parks department or tree officer.

• Local tree surgeons may be able to deliver small amounts, but they are unlikely to want to sell full tree lengths (most tree surgeons move timber around by hand).

- The Myforest online tool also contains a searchable map of timber sales.

- Forestry contractors may have small lots that they are interested in selling.

- Saw mills often have surplus 'slab wood' (semi-circular off-cuts) that they are interested in getting rid of.

- Local farms and estates may have available timber.

FELLING AND PROCESSING

If you own a piece of woodland or have forests locally that can supply you wood in lengths (before cross cutting) then this will probably be cheaper than buying processed logs, though you will need to buy an axe (or splitting maul) or hire a log splitter and spend a lot of time processing your wood. The BEC has a section on their website about managing a woodland for fuel (see Appendix 3 for details) and the HSE provides a series of guides on how to conduct forest operations safely. At this point you're getting beyond the scope of this book. I have included a few links to useful books on this topic at www.wrolls.co.uk links if you want to read up further.

Do watch out if you're planning to fell trees on your land that you're not falling foul of any legislation. The assumption is always that you will need a felling licence unless the tree is growing on land that is for some reason exempt, and even if you're confident that you don't need a felling licence, you may find that you are in a conservation area or that there's a European Protected Species on the site. The safest way

87

through this minefield is to contact your local woodland officer at the Forestry Commission. All the contact details are on their website at www.forestry.gov.uk

If you're processing your own wood, you may choose to cross-cut by hand or with a chainsaw (in which case a measuring stick is really useful) or hire in a log processor. These processors cut to length automatically and can be calibrated to whatever size is appropriate. They will usually also split the log into a convenient size. Beware if you have a lot of large diameter trees (broadleaves particularly) because you may exceed the maximum in-feed diameter of the processing unit (anything under about 8 inches should be fine). There are bits of kit that will split whole tree lengths to fit in firewood processors, but they will be significantly more expensive and will certainly require an experienced operator to work the machinery safely. You may also run into difficulties if your timber is not reasonably straight. (There's not much point in being able to cross-cut automatically if you have to do it manually before the log will fit in the processor.)

■ A word on chainsaws

If you are using a chainsaw, **you really do need to get some training and the proper protective gear**. Chainsaws are about the most dangerous bit of kit you can buy in this country without a licence. **They really do want to eat you**, and will try at every opportunity. It's not good to be on your own in the woods (or anywhere else) if you've just taken a chunk out of your leg. To put things in perspective, according to the HSE guidance on chainsaws:

'Anyone working with chainsaws needs to understand how

to control major bleeding and to deal with crush injuries...'

Though I suspect that if you had injured yourself, you wouldn't be in much condition to do either.

According to the Royal Society for the Prevention of Accidents,* over a thousand people each year are injured by chainsaws outside of a work environment. **You need to take safe handling and proper maintenance seriously.**

Lantra (the sector skills council for land management and environmental skills training) is probably the best place to look for courses on chainsaw use. They have a list of courses online here: www.lantra.co.uk/CourseFinder.aspx

TRANSPORTATION

Logs don't stack neatly and often contain a significant amount of water. Transportation of this sort of material can be expensive. Often it's cheapest to transport logs as tree lengths as it's easiest to load and unload with a hydraulic grab. Bear in mind that even if you can get logs delivered, you may still need to move and stack them and you may not be very popular if they are getting in everyone's way. This is an even more significant issue if you've had it delivered in two-metre lengths.

DRYING

Assuming you aren't able (or don't need) to find a supplier of dried logs, how do you turn green logs into dry logs?

---

* www.rospa.com

There isn't a big secret. What you need most is the space to stack them, and the time to let them dry. All you need to do is increase the size of your existing log pile, or build a new one (see the section on storing fuel above.) Felling one winter to burn the next is common; if you have the space, leaving the wood for two summers is better. After this, the benefit of further drying decreases, as the natural humidity in the air makes it very difficult to get logs down to below about 20 per cent MC in the UK without some element of forced drying.

Logs dry in proportion to their volume and surface area. Short logs dry more quickly than long logs, and small diameter logs will dry faster than logs with larger diameters. Try to make sure that most (if not all) your logs have a split face. This greatly increases the speed of drying as bark forms a water resistant layer around the log which slows down the rate at which moisture is able to evaporate. Radial cracks and bark that peels off easily are usually good indicators of dry logs.

Appendices

I. CALCULATIONS AND CONVERSIONS (THE *REALLY* GEEKY BIT)

Area

Nearly all calculations of yield are based on metric units, which means hectares (ha) rather than acres.

1 acre = 0.40 ha and 1 ha = 2.47 acres

Moisture content

■ Changes to weight with drying

For some reason, this calculation seems to confuse a lot of people. However, it's a really useful method to work out what weight of fuel you'll be left with when drying has taken place.

$$\frac{W_1 * (1 - MC_1)}{(1 - MC_2)} = W_2$$

W_1 Green Weight
W_2 Dried Weight
MC_1 Green Moisture Content (wet basis, expressed as a decimal e.g. 50 per cent = 0.5)
MC_2 Dried (target) Moisture Content (wet basis, expressed as a decimal e.g. 50 per cent = 0.5)

91

For example: 1 tonne of wood at 50 per cent moisture contains 500 kg of water and 500 kg of dry matter. If you dry the wood to 30 per cent moisture content, the dry matter weight remains unchanged, but instead of forming 50 per cent of the total weight, it now forms 70 per cent. To find out what the 500 kg is 70 per cent of, divide by 0.7 (the result is 714 kg).

In case you were wondering, there is no reason why W_1 above has to be greater than W_2, or MC_1 greater than MC_2. It is therefore possible to use the same formula to work out how many tonnes of green wood you need to generate a particular number of tonnes of fuel. Simply switch W_1 and W_2, and MC_1 and MC_2.

- ■ *Wet basis and dry basis measurements*

Wet basis is the proportion of a given sample that is water. If I hold a log in my hand, the amount of water within that wood is a proportion of the total weight. That is the wet basis.

$$\frac{W}{T} * 100 = MC\%_{wb}$$

W	Weight of water in a sample
T	Total weight of the sample
$MC\%_{wb}$	Moisture Content (in % wet basis)

Dry basis is the proportion of water in a sample as expressed against the oven-dry weight of the wood. The advantage of this is that the dry matter proportion of the wood remains constant. This is used by professionals who need to calculate the load bearing capacity of wood (water doesn't contribute to the structural properties, only the weight.) It's rare that you'll come across fuel sellers using dry basis moisture content.

$$\frac{W}{O} * 100 = MC\%_{db}$$

W Weight of water in a sample
O Oven-dry weight of the sample
$MC\%_{db}$ Moisture Content (in % dry basis).

- ## Converting wet to dry basis (and back again)

$$\frac{MC_{wb}}{(1 - MC_{wb})} = MC_{db}$$

$$\frac{MC_{db}}{(1 + MC_{db})} = MC_{wb}$$

MC_{wb} Wet basis moisture content expressed as a decimal (i.e. 50 per cent = 0.5, etc.)
MC_{db} Dry basis moisture content expressed as a decimal (i.e. 50 per cent = 0.5, etc.)

- ## What basis is that?

If someone doesn't tell you what type of moisture content they are talking about and they are selling you fuel, they are almost certainly talking about wet basis, but you can look very smart (or annoying) by asking them to double check. You can get moisture meters that will give you a reading in either format and they should say on them which one they measure. If you happen to get hold of a moisture meter that doesn't say (or you're not sure) the easiest way to check is to find a really wet piece of wood and test it. If the reading is over 100% then you are looking at a dry basis measurement (50% wet basis = 100% dry basis).

Estimating volume

▪ Standing volume

There is a really useful tool for getting a ballpark figure for standing volume on the BEC website's Technical Development Reports* and Tools section. It was written by (among others) by Robert Matthews and Ewan Mackie who are the authors of the (in)famous Forestry Commission 'Blue Book'.†

▪ Felled volume

It is much easier to measure trees when they have been felled and had branches cut off, etc. Normally I would suggest measuring the stack size and then multiplying by a conversion factor to take account of the air space. The usual factors will tend to be somewhere between 1.5 and 2.5 depending on how neatly the material is stacked. Again, there is a detailed methodology in the Blue Book.

Energy

▪ GJ per tonne

Most oven-dry timber has a net calorific value of about 19 GJ per tonne. There is variation between species, but since there is so much variation between trees of the same species, and even different parts of the same tree, an approximate figure is good enough. The 2.443 figure is the energy necessary to boil

---

* www.biomassenergycentre.org.uk/portal/page?_pageid=74,373197&_
dad=portal&_schema=PORTAL
† Matthews, R and Mackie, E, *Forest Mensuration: A handbook for practitioners*,
Forestry Commission, Edinburgh, 2006.

away the water in the fuel (the latent heat of evaporation).

$$(19 * (1 - MC_{wb})) - (2.443 * MC_{wb}) = GJ / tonne$$

MC_{wb} Wet basis moisture content expressed as a decimal
(i.e. 50 per cent = 0.5, etc.)

■ GJ to kWh

3.6 GJ = 1 MWh
1 MWh = 1,000 kWh

■ Old units

I have to admit that I hardly ever encounter a btu (British Thermal Unit) or a therm in the wild, but if you do:

1 btu = 1.055 kJ
1,000 btu = 1.055 MJ
100,000 btu = 1 therm or 105.5 MJ

2: TROUBLESHOOTING

The fire won't catch or light properly

The wood needs to be dry, small diameter, with a good air supply. This problem can be caused by material that is too thick or damp, or (rarely) if you haven't got enough air reaching the fire.

The wood smoulders and won't burn with a bright flame

This sounds like a problem of insufficient secondary air – you may be starving the fire of oxygen. Try opening the draught to see if that helps. If you still run into difficulties you may need to get the chimney swept or attach a draught regulator to the flue.

If there's a lot of steam and the wood hisses as it burns, you've probably got wet fuel. Make sure it's been dried properly.

Lots of soot is forming. The stove walls and window (if present) don't stay clean

This is likely to be caused by a stove that is not burning hot enough.

The wood may be too damp. Make sure it's been dried properly.

You may be starving the fire of oxygen (secondary air) try opening the draught to see if that helps.

If the stove has a back boiler, the water jacket may be cooling the firebox leading to increased condensation. You should have a return plumbed in to prevent cold water cycling through the heat exchanger. Check with an installer to see whether this is the case. Otherwise, the only solution I know of is to add a lot of dry wood and try to heat the stove out of the condensation zone.

Although the fire burns strongly, the stove isn't radiating much heat

You may be drawing too much air into the system. Try closing the draught slightly to slow the passage of gasses through the system. If the problem persists, you may need to install a draught regulator on the flue.

Wood burns too quickly with low heat output

You may be drawing too much air into the system. Try closing the draught slightly. If the problem persists, you may need to install a draught regulator on the flue.

The fuel may have been cut up too small – try adding a larger log – or may (very occasionally) not be dense enough. Try adding some denser fuel to see if it makes a difference.

You may just be using the stove in a way it wasn't designed to operate. Check the instructions to see if there's anything you missed (we've all done it!).

Smoke escapes through the stove door

This is often a problem caused by the chimney failing to draw properly. Ideally you want a strong convection current pulling air up the chimney.

You may not be adding enough air into the system. Try adding more secondary air.

The chimney may be blocked. Try getting it swept. If this doesn't work, you may need to take more drastic action like getting the chimney widened, or adding a chimney cowl to prevent downdraughts.

In addition to these problems, you may have a poor seal between the stove door and the room. Make sure you've sealed the door tightly. Check the fire rope around the door – it may need replacing.

The chimney becomes moist and sooty; condensation drips down from the chimney

This could be caused by wet fuel, or running the stove too cool. Remember you need the flue gases to get out of the top of the chimney before they drop below 120°C.

This may also be caused by a chimney that is too wide. If the problem persists you may need to get it looked at to see if you need to add a smaller diameter liner.

Smoke escapes into other rooms

Your chimney may need re-lining (or you may not even have a liner fitted). **Stop using the stove until you can get it checked out by a qualified engineer.**

3: WHERE TO GO FOR FURTHER INFORMATION

I have included a number of references to useful resources throughout the text, some of which are available for free online, or from good bookshops. Unfortunately most of them have irritatingly complicated internet addresses, which would be a real pain to type in from hard copy. Rather than expecting you to type in all the addresses by hand, I have made a list of all the resources available online (www.wrolls. co.uk/thelogbook2/reference) for you to access directly.

4: FEEDBACK

I know that there are a lot of different people who spend a lot of time using woodburners, and I expect that some of you will think I've got something wrong in this book. I'd love to get feedback on the points where you have different ideas, or things you think I could have explained more clearly. If it turns out I've made any grievous errors, I'll make sure I post them on my website, so you can all have a good laugh. I'd welcome more general feedback from readers too – you don't have to be a stove expert!

You can tell me all about it at www.wrolls.co.uk/contact; or via email to feedback@wrolls.co.uk; or on the book's Facebook page: www.facebook.com/BurningWood

Glossary

Biodiversity
A measurement of the diversity of species within a given ecosystem. A more biodiverse habitat will support a greater number of different species. While this is not precisely equivalent to a measurement of habitat value (some rare habitats for endangered species are not particularly biodiverse), it is usually a fairly reliable indicator of habitat value (in the UK) when compared to habitats that have suffered degradation from human interaction.

Boiler
A combustion unit that produces heat to be used away from the site of combustion, usually via piped water and radiators.

Calorific value
The useful energy released when burning a given weight (mass) or quantity (volume) of a fuel.

Carbon dioxide (chemical formula CO_2)
A colourless, odourless gas that is the product of complete combustion. CO_2 is non-toxic, but is a cause of anthropogenic (man-made) climate change via the greenhouse effect.

Carbon monoxide (chemical formula CO)
A colourless, odourless gas that is a product of incomplete combustion. Carbon monoxide is poisonous when inhaled in sufficient quantities and is a hazard to health. Alarms are available commercially which detect the presence of CO in a room, and these are strongly recommended when using log stoves.

Carbon-lean
A fuel type which, while not being completely carbon neutral

(no net CO_2 emission), has a very small net emission. Wood is usually referred to as carbon lean, as small amounts of fossil fuel are commonly used in felling, processing and transportation.

Deforestation
Conversion of forest land to other purposes by the removal of trees. This is an extreme example of unsustainable forestry, which can usually be defined as the removal of timber at a rate exceeding the growth of trees on the site.

Dioxin
Dioxins and dioxin-like compounds are by-products of various industrial processes, and are commonly regarded as highly toxic compounds that are persistent environmental pollutants.

Endothermic
A chemical reaction that absorbs heat from the surrounding environment.

Exothermic
A chemical reaction that emits heat.

Fossil/contemporary carbon
Carbon that is active in the current carbon cycle is described as contemporary. Living organisms and atmospheric carbon are examples of contemporary carbon. Fossil carbon has been removed from the contemporary carbon cycle by sequestration into an inert state (for example fossil fuels). Burning fossil fuels transfers fossil carbon into the contemporary carbon cycle – this is the root cause of the greenhouse effect and anthropogenic climate change.

Fossil fuels
Materials that contain carbon that has been 'fossilised' or sequestered millions of years ago. Includes natural gas, oil and coal.

Heating season
The period of time over which additional heating is required in a property. It depends on the weather, location, and how much warmth the owners require, but usually from about October through to about the end of February.

Heavy metals
A loosely defined range of elements, including lead, arsenic, mercury, chromium and cadmium, that exhibit metallic properties. They are have a toxic effect on health and the environment. This effect can be exacerbated by their tendency to accumulate within the food chain.

Incomplete combustion
Combustion where insufficient oxygen is present to completely react with the fuel. This leads to a number of reliability problems and potential health hazards.

Joules (J)
A unit of energy equal to the force required to move one Newton the distance of one meter.

Kilowatt hour (kWh)
A unit of energy equivalent to 3.6 mega joules (3,600 J).

Lignin
Lignin is a chemical compound found in woody plants. Its most commonly recognised function is to add rigidity.

Particulate
Tiny quantities of solid matter that are suspended in the atmosphere. Particulates come from a number of sources including diesel vehicles, log stoves, and a range of natural sources. They are generally referred to as PM_{10} and $PM_{2.5}$ (Particulate Matter 10 microns or less, and 2.5 microns or less.) The figures of 10 microns and 2.5 microns relate to the effects that these particles have on the human body. High concentrations of particulate matter can cause lung complaints.

Photosynthesis
The process used by green plants of absorbing sunlight, carbon dioxide and water to produce glucose and oxygen.

Primary and secondary air
Air flows feeding combustion in different locations. Primary air is directed at the fire bed, and secondary air is directed at the burning gaseous compounds above the fire bed. See page 22 for primary and secondary combination reference.

Primary and secondary combustion
Primary combustion is burning of the solid char component of a fuel, secondary combustion is burning of released volatile gases released by wood as it is heated. See page 10 for the full definitions.

Sequestration
The process of removing carbon from the atmosphere and depositing it in a carbon sink or reservoir. Photosynthesis is one example of sequestration.

Stoichiometric ratio
An optimal mix of reactants to achieve a complete reaction with no reactants remaining at the end. When looking at combustion this relates to a complete combustion of fuel and air with no products of incomplete combustion remaining.

Stove
A combustion unit that heats the room it is in. Hybrids between stoves and boilers do exist, which heat the room the unit is installed in as well as providing domestic hot water and some limited radiator capacity.

Wet basis and dry basis moisture content
See page 92 for the full definitions.

Index

About the author

Will Rolls is a chartered forester who specialises in the use of wood as fuel. Until recently, he worked for the Biomass Energy Centre, a small unit within the research agency of the Forestry Commission, where he spent most of his time giving technical advice on the availability, use and development of woodfuel and wood fired heating systems. He now operates an independent woodfuel consultancy advising clients on their potential fuel requirements and supply capability. You can find out more at www.wrolls.co.uk

Will lives in Yorkshire with his wife and daughter. If you give him a woodland to play in, and a stick to poke things with, he'll be a happy man.